UNIVERSE IN CREATION

UNIVERSE IN CREATION

A New Understanding of
the Big Bang
and the Emergence of Life

Roy R. Gould

Harvard University Press

Cambridge, Massachusetts
London, England
2018

Library of Congress Cataloging-in-Publication Data

Names: Gould, Roy, 1947– author.
Title: Universe in creation : a new understanding of the big bang
and the emergence of life / Roy R. Gould.
Description: Cambridge, Massachusetts : Harvard University Press, 2018. |
Includes bibliographical references and index.
Identifiers: LCCN 2017047932 | ISBN 9780674976078 (alk. paper)
Subjects: LCSH: Universe. | Life—Origin. | Creation. |
Big bang theory. | Beginning.
Classification: LCC QH325 .G68 2018 | DDC 570.1—dc23
LC record available at https://lccn.loc.gov/2017047932

To Leora

Contents

Introduction   1

Part One
WHERE DOES THE UNIVERSE COME FROM?

1. What Is the Universe—and How Large Is It?   19
2. Galaxies Misbehave   37
3. What's the Big Idea?   43
4. Einstein, Gravity, and the Universe   51
5. The Big Bang and Beyond   70
6. Building Plans   83

Part Two

HOW DID STRUCTURE ARISE FROM CHAOS?

7. An Apple Pie from Scratch 95

8. Into the Abyss 105

9. Into the Cauldron 111

10. Into the Light 124

Part Three

IS LIFE MERELY A ROLL OF THE COSMIC DICE?

11. The Great Inventor 143

12. Information, Please! 164

13. Is Evolution Predictable? 185

14. The Sensational Sensations 204

15. Design without a Designer? 213

16. "Who's There?" 222

Epilogue: What Is Worthy of Our Wonder? 239

Notes   247

Acknowledgments   267

Index   269

UNIVERSE IN CREATION

Introduction

In just the twinkling of an eye—a mere three hundred million years from now—the universe will turn fourteen billion years old. When you get that old, you run the risk of becoming a mere compilation of facts, a mind-numbing history. But does the universe also have a genuine story to tell? One with a dramatic arc? A story with a point to it?

Scientists don't normally treat nature as though it were a character in a novel, replete with motives and intentions and purpose. After all, science concerns itself with what is testable, at least in principle. Think how difficult it is to know another person's motivations; now imagine how much more difficult and presumptuous it would be to decipher the "intent" of a silent universe.

But the universe has not wandered aimlessly for those billions of years. It has been at work, hard at work, and it has unfolded with a logic and beauty that stagger the imagination. We now know so much about the universe—its origins and infancy, its growth, and even its relationship to life— that we are in a position to make sense of what the universe has been doing and what its story might be.

Charles Darwin was quick to point out the need to find meaning among the facts. After he published his *Origin of Species,* he received an irate letter from a scientist urging him to stick to the facts and to leave out any presumption of a story. "It made me laugh," Darwin wrote to a friend, "to read of his advice or rather regret that I had not published facts alone. How profoundly ignorant he must be of the very soul of observation. . . . At this rate a man might as well go into a gravel-pit & count the pebbles & describe their colours. How odd it is that every one should not see that all observation must be for or against some view, if it is to be of any service."[1]

The view of the universe we are about to explore is likely to be very different from the one to which you are accustomed. A wealth of research—from fields as diverse as astrophysics and zoology—is helping to paint a new portrait of the universe and to address deep questions about the cosmos that would have been off the table just a few years ago. To get the most out of this adventure, first we need to examine our preconceptions.

A Universe Still in Creation

If you are like many people, you may assume that the creation of the universe took place long ago and far away, in a dramatic event called the Big Bang. Out of reach, out of sight, out of mind—and inscrutable except to an Einstein or a deity. And you might also assume that after the Big Bang, the universe was simply left adrift to fend for itself. Perhaps you think of the universe as you might regard a giant sports arena: a vast, empty space where events show up from time to time and then leave the stage. Or where nothing much at all happens. You may think of the universe as a backdrop, not an actor.

So it may come as a surprise that the Big Bang was *not* the creation of the universe. Yes, matter did spring into existence back then. But it wouldn't have done you much good: the entire universe that we observe today was, back then, smaller than you are now; the space you occupy hadn't yet been created!

The creation of the universe is an ongoing process, not an event. It is taking place even today. Out in deep space, between the galaxies of stars, new space is continuously streaming into existence, swelling the volume of the universe. In fact, most of space was created after life had already arisen on Earth. So too is new energy pouring into existence; the majority of energy in the universe also was created long after the Big Bang. New stars are forming, chemical elements are being manufactured, and new worlds and presumably new life are being added to the universe's résumé as you read this. This is a universe at work.

The very term "creation *of* the universe" is misleading. It suggests that some external force shaped the universe, like a piece of pottery on a wheel. But we will see that the universe itself is the master of creation—what Shakespeare called "great creating nature." The universe is able to generate more of itself, even as it creates everything *in* itself.

Like us, the universe is a work in progress.

Does the Universe Have a "Plan"?

Just as you might look over the shoulder of an artist at work and try to figure out what the artist is trying to accomplish, so too can we look at our universe in progress, sift through what we know about it, and try to deduce the bigger picture, to tell not just how the universe works in its minute details but what, if anything, it is trying to do.

You might think that the universe could not possibly have a plan. The word *plan* seems to imply intent—and intent is something we ascribe only to living things, such as us and our golden retrievers. A rock doesn't have a plan, so how could the universe? In fact, the whole idea at first seems so unscientific that you might suppose it defines the boundary between science and the occult. "The scientific attitude," said the biologist and Nobel laureate Jacques Monod, "implies what I call the postulate of objectivity . . . that there is no plan, that there is no intention in the universe."[2]

So you may be surprised to learn that the universe does indeed come with a plan. The plan is the *infrastructure* with which the universe was born. This birthright includes the basic particles of matter, as well as the forces that animate these particles and set them in motion on the cosmic stage. It also includes an orderly set of rules that guide the universe as it unfolds from moment to moment. A physicist might call these rules the "laws of nature," but of course our knowledge of them is only approximate, to be continually updated as we learn more.

In this book, I'll refer to the infrastructure of the universe as its *building plan*. The building plan determines some things completely and leaves others completely undetermined. In fact, chance and uncertainty are built into parts of the plan itself, such as in the quantum behavior of the microscopic world. Deciphering the plan, and understanding its consequences, is the subject of virtually all of the physical sciences. So the question is not whether there is a plan. The question is, *What is the plan and what are its consequences?* We will return to this question throughout the book.

Back in 1884, when Mark Twain published *The Adventures of Huckleberry Finn,* no one had much of a grasp of the universe's building plan. "We had the sky up there, all speckled with stars," said the novel's young adventurer, "and we used to lay on our backs and look up at them, and discuss about whether they was made, or only just happened. Jim he allowed they was made, but I allowed they happened; I judged it would have took too long to make so many."

Today we know that the billion trillion stars that grace our universe didn't "only just happen"; they followed from the universe's building plan. Neither is a planet like Earth an accident; there are several billion planets roughly the size of Earth in our Milky Way galaxy alone. On the other hand, if you want to know what the weather in London will be a million years from now, you will just have to wait. Nature has no idea, any more than it knows what you're going to have for lunch tomorrow. That is not written into the plan!

The Universe Is an Entity in Itself

I am sure that you are still uncomfortable with the idea that the universe has a plan, that it is going somewhere, trying to accomplish something as it unfolds. And why shouldn't you be skeptical? It is not a comfortable idea.

For one thing, a casual glance at the night sky suggests that the universe is idle. The romantic tableau of stars that you saw the other night is virtually identical to the night sky that your ancestors saw thousands of years ago. And isn't that where the night's romance comes from? It is deep and dark and unchanging, far removed from the hubbub of our world. Even Einstein originally assumed that the universe remains the same "from everlasting to everlasting."

Then too there is the fear that the universe has a plan for *us*. Just consider that the stars were once accorded the power to determine our destinies, as the "star-crossed lovers" Romeo and Juliet found out the hard way. The very word *disaster* literally means "ill-starred." Shakespeare had to remind his audiences that "the fault . . . lies not in the stars, but in ourselves." Even the archvillain Edmund, in *King Lear,* assures us that he would be just as evil "had the maidenliest star in the firmament twinkled on my bastardizing." It has taken centuries to clear the mind of superstitious cobwebs, and surely we are wary of anything that might cloud it once again.

Furthermore, we seem to be more intrigued by the universe's mysterious origin than by its forward motion. We imagine that the deepest thing we can know about the universe is where it came from. Physicists have a special love for the deep past, because the universe was much simpler in its infancy than it is today. If you want to uncover the basic laws of physics, it helps to look back in time as far as you can. Today, for example, the various forces of nature, such as gravity or magnetism, behave very differently from each other, and they require very different descriptions. But when the universe was young and much hotter, the forces resembled each other much more closely; it is as though there was only one simple force. This is one of the reasons that physicists have spent billions of dollars to build enormous atom smashers that can re-create the conditions that existed a trillionth of a second after the Big Bang. The hope is to produce for study the very simplest particles and interactions that underlie the universe.

Despite the lure of origins, I hope to convince you that the forward motion of the universe is at least as profound as its origins. This unfolding of the universe—its story—may help us decipher *why* the universe is built the way it is.

Finally, we need to confront a remarkable misconception about the universe that most of us have: we think of the universe as the sum of everything there is—a vast collection of space, stars, planets, and so forth. That is the dictionary definition, and it seems reasonable. But it turns out not to be an accurate depiction of the universe.

The universe is much more than the sum of its parts, just as you and I are much more than a list of the molecules from which we are constructed. The universe is an entity unto itself, different from anything else we know. For example, while the universe has laid down the physical laws that all of us must follow, it has nevertheless found loopholes that apply only to itself. We'll see that the universe-as-entity can create new space; the rest of us cannot. The universe-as-entity can bring new energy into existence; the rest of us can only transform energy from one kind to another. You and I cannot exceed the speed of light—yet the expanding universe sends galaxies hurtling apart faster than light. And, to add to the wonder, we'll see that the universe is expanding inward, not outward— a mind-twisting feat that only the universe can accomplish! As we'll see, the universe is neither animate nor inanimate. We don't yet even have the language to describe the universe properly.

All of this should alert us to keep our minds open, to examine our preconceptions about what the universe can and cannot do, what it is and is not.

Is Life Written into the Universe's Building Plan?

In 1983, the physicist John Archibald Wheeler posed a simple question that we will return to repeatedly in this book. Wheeler spent much of his professional life at Princeton University, where his research ranged from the subatomic world, to the physics of the universe, to black holes (a term that he coined). But his question transcends any one field of science: "Is the machinery of the universe so set up," he asked, "and from the very beginning, that it is guaranteed to produce intelligent life at some long-distant point in its history-to-be?"[3]

Put simply: Were we on the universe's to-do list?

At first glance, the question might seem irrelevant. We live our lives perfectly well from cradle to grave without worrying in the slightest about whether the universe "wants" us here. What difference would it make? But Wheeler's question is not about us; it is about the universe. If you want to fully understand the universe and what it is up to, then eventually you must address his question. Otherwise, it would be like studying the punctuation marks in a novel but having no idea of the plot.

In the era when Wheeler posed his question, most scientists assumed that the universe has no stake in the emergence of life, and many scientists still feel this way. The words often used to describe the emergence of life are "chance," "improbable," and "accident." Life is seen as a kind of epiphenomenon, like a guest who showed up for dinner but wasn't invited.

Jacques Monod put it bluntly: "The universe was not pregnant with life, nor the biosphere with man." He continued, "Chance alone is at the source of every innovation, and of all creation in the biosphere. Pure chance, absolutely free but blind."[4]

The evolutionary biologist Stephen Jay Gould wrote, "We are the accidental result of an unplanned process . . . the fragile result of an enormous concatenation of improbabilities, not the predictable product of any definite process."[5]

The Nobel Prize–winning physicist Steven Weinberg added a psychological twist. "It is almost irresistible," he wrote, "for humans to believe that we have some special relation to the universe, that human life is not just a more-or-less farcical outcome of a chain of accidents reaching back to the first three minutes, but that we were somehow built in from the beginning."[6]

But Wheeler's question was motivated by the intuition that we *do* have a special relation to the universe—that intelligent, conscious beings are in some sense necessary to the universe's existence. This intuition came not from theology or biology but instead from deep puzzles in physics. For example, the *observer* plays a fundamental role both in quantum physics and in the perception and measurement of time. Many of the equations and ideas of physics have meaning only in terms of what someone observes or measures. Wheeler called it a participatory universe.

As the cosmologist Andrei Linde put it, "The universe and the observer exist as a pair. You can say that the universe is there only when there is an observer who can say, Yes, I see the universe there. . . . The moment you say that the universe exists without any observers, I cannot make any sense out of that. I cannot imagine a consistent theory of everything that ignores consciousness."[7]

It is a subtle point. Most of us would say that the universe would exist even if we had never been born. But we come to that conclusion only because we *were* born.

The universe's building plan has a remarkable property that provides support for Wheeler's and Linde's conjecture: the plan appears to be "fine-tuned" for life. If you tried to tweak some of the thirty-odd parameters of the building plan significantly, then you would alter the universe in ways that preclude life. For example, if you could make gravity stronger than we observe it to be, then stars might burn out too quickly for life to arise or they might not form at all. A mutant universe doesn't do anything interesting.

Why should the universe have been born with a building plan so hospitable to life? Obviously, if the laws of nature had precluded life, then we wouldn't be here to ask the question. But physicists have not been able to deduce, from established principles of physics, why the building plan must be the way it is. Many physicists are now considering the possibility that there are a vast number of universes beyond our own, each with its own building plan. Some theories predict that there are many ways to construct a universe and that most of these alternative building plans would preclude life as we know it. The theories that describe other universes are highly speculative, but they remain attractive to many scientists, and they provide a possible explanation for why our universe is constructed so benevolently: we are simply lucky.[8]

Perhaps, the speculation goes, our own universe's building plan is based on chance and reflects nothing more than a roll of the cosmic dice. According to Tim Maudlin, a physicist and professor of philosophy at New York University, "Our modern understanding of cosmology demote[s] . . . the very existence of our species to mere cosmic accident. . . . In the end, we might just have to accommodate ourselves to being yet another accident in an accidental universe."[9]

Not all scientists agree.[10] The biologist and Nobel laureate Christian de Duve saw the appeal to multiple universes as the latest manifestation of what he called "the gospel of contingency"—only now it was the universe rather than us that was said to arise by chance. De Duve considered life and mind "such extraordinary manifestations of matter" that they remain meaningful regardless of how many universes exist or are possible.[11]

Then which is it? Has life been on the universe's agenda all these years, as Wheeler conjectured? Or is life merely a happenstance? A roll of the cosmic dice?

Wheeler was not a fan of theories of multiple universes, which he called "outside the legitimate bounds of logical discourse."[12] It was enough to try to understand our own universe, without invoking realms forever beyond observation. But he also had no illusions about how difficult his own question would be to answer.

Wheeler's question is not what scientists call well posed. After all, it is a big universe out there, with a lot of time on its hands. Where is life "guaranteed" to arise, and under what conditions? How long does life take to emerge? Does it show up just once or twice, or is the universe teeming with life? And suppose that we were the only life-form in the universe. Would that satisfy the guarantee? How could we determine the difference between being on the universe's agenda and being merely accidental?

Wheeler could not comprehend how a universe that started in chaos could bootstrap its way to something as complicated and wondrous as intelligent life. What sequence of mechanisms would guarantee life? How could the universe know where it is going and what it needs to do to get there?

"How should such a fantastic correlation come about between big and small," he wondered, "between machinery and life, between future and past? . . . In brief, how can the machinery of the universe ever be imagined to get set up at the very beginning so as to produce man now? Impossible! Or impossible unless somehow . . ."[13]

This book explores the "somehow": the scientific evidence that bears on Wheeler's question. Although we still don't know the answer for certain, the available evidence suggests that the universe may well be set up to guarantee the emergence of intelligent life. In the law, a prima facie case is one in which the facts warrant further discussion; this book is intended to begin just such a discussion.

I will take two approaches to Wheeler's question. One is to explore the story of the universe as it has unfolded from the Big

Bang to the present day. What can we deduce from the story? In what ways has nature put out a welcome mat for life? We will find that ever since the moment of the Big Bang, the universe's major construction projects have laid the groundwork for life in an astonishing variety of ways.

For example, you and I and the world's greatest engineers could not accomplish very much using only the simplest element, hydrogen. Yet the universe has fashioned this unassuming material into the great engines of creation—the stars—which manufacture chemical elements and spew them across space. At the same time, stars are able to warm and illuminate their planets without fail for the billions of years that it takes for intelligent life to evolve. The stars are self-assembling, complex machines that are essential to life, and they fill our universe thanks to the building plan with which the universe was born.

But the groundwork for life is not the same as life itself. We still cannot prove that life follows inevitably from the building plan. Life is too complex, and the path to its origin is too far removed from us in time. So we will take a second approach. We'll examine various properties of life to see what they might plausibly indicate about Wheeler's conjecture.

Regardless of the answer to Wheeler's question, our view of the universe and our place in it is about to receive a jolt. The world has now embarked on a great adventure: the targeted search for life in our galaxy (see Chapter 16). When I was a child, the search for life meant only the search for extraterrestrial intelligence. It was a noble quest with long odds. No one knew where to look, since the only known planets were the ones in our own solar system. No one knew when to look, since there was no way to know when an alien was feeling communicative. And no one really knew how to look, although radio waves seemed a good idea at the time.

Today, thousands of planets have been discovered. Their celestial zip codes have been mapped and their portraits are beginning to be sketched. Dozens of these worlds are in the habitable zones of their stars, where liquid water and therefore life are possibilities. Billions more planets in the Milky Way galaxy await discovery. Best of all, every one of the known planets is "speaking" right now, sending to Earth a continuous stream of light that has bounced off the planet's surface or refracted through its atmosphere, waiting to be decoded. This light can speak volumes about the planet's atmosphere, continents, climate, and evidence of life. Astronomers have already begun to analyze the atmospheres of some of these worlds—and even to construct the rudiments of a weather report for alien worlds trillions of miles away!

After years of wondering who or what inhabits the great sea of space and time, we are finally poised to find out.

What about God?

In this book, I take the poet Walt Whitman's advice to "make no arguments concerning God." But that doesn't mean that *you* can't. People bring their own backgrounds and understandings to any discussion about the universe. A physics teacher I once worked with told me that every time he saw the latest image from NASA's Hubble Space Telescope, he felt that he was "looking at the face of God." The sight of stars forming from majestic clouds of dust and gas was like being present at the Creation, he said.

But another teacher had the opposite reaction. "Where is God?" she pressed. "You've been to the Moon, seen the surface of Mars, seen all the way across the universe. So where is God?" She had grown up in a remote village in Africa, where she and her sister were tormented by villagers who told them that the gods were angry every time a comet or other celestial omen appeared in the night sky. She said that it was a relief to look past the Moon, past Mars, and all the way out, and see a vast and remote space that was blissfully devoid of any deities.

Scientists often use the word *God* as though it were synonymous with nature, such as when they ask, "Why did God create the universe *this* way, instead of some other way?" And they often use the word *God* mischievously. This was brought home to me a number of years ago when I had lunch with the British physicist Stephen Hawking on his visit to the Museum of Science in Boston, along with a young physicist and a nurse who traveled with him. Hawking uses a wheelchair and speaks by means of a keyboard-controlled robotic voice.

I happened to mention that the museum was seeking funding from the corporate giant IBM for an exhibition on the Big Bang. Hawking suddenly grew visibly distressed. He typed out a few words on his keyboard. Suddenly, the robotic voice squawked over and over again, "IBM doesn't love God! IBM doesn't love God!" Hawking had an impish grin on his face, indicating some wry humor was at work.

It turned out that IBM had agreed to sponsor a documentary film about the universe that he was working on, provided that no one in the film mention the word *God*. That would have been very difficult for all the physicists who invoke God without even thinking about it!

Mechanism, Mystery, and Miracle

Science deals largely with the mechanisms of how things work. Although we will certainly explore how nature works, this book focuses on how we might make sense of what we find out about nature. In particular, this book highlights what is miraculous about the unfolding universe.

By "miraculous," I don't mean that it involves divine intervention or the suspension of natural law, as in the parting of the Red Sea. That kind of miracle makes for good stories but terrible history. And I don't mean something that is so mysterious or complex that "it could only be a miracle." That is more like a magician's trick. Once you see how it is done, the effect is ruined.

To me, the true miracles of nature are just the opposite. They are the things that we have come to understand and explain yet that seem so beautiful, so right, so clever, that you can only nod in enjoyment or smile an inner smile. We have all seen this kind of miracle in an athlete who defies what we thought was possible, or in an artist who brings into the world some simple beauty. You can see how it is done, yet you still can't believe that it is actually done.

Nature is the ultimate miracle maker, truly worthy of wonder, as I hope you'll discover throughout this book.

Part One

WHERE DOES THE UNIVERSE COME FROM?

1

What Is the Universe—and How Large Is It?

Today you can simply Google the size of the universe and let the numbers magically appear from the cyberworld and wash over you. But if you really want to understand the scale of the universe and what it means, it helps to remember that our understanding of the universe is the result of a long series of steps, spanning centuries, each step allowing us to peer farther out into the depths of space. And it may seem strange to us today, when answers come so easily, that people once sacrificed their lives and livelihoods to take the measure of the universe.

Two episodes in that long chain of understanding will reveal a great deal about the immensity of the cosmos. But they are equally important because of what they reveal about us as a species, and about our drive to fathom our place in the universe.

How Far Away and How Large Is the Sun?

In Greek mythology, the god Helios carried the Sun across the sky each day in his chariot. Helios's young son Phaeton secretly took the chariot for a joyride, but, as you might guess, it didn't end well. Phaeton lost control of the horses and plummeted from the sky, scorching the Sahara desert and burning himself in the process. Zeus was forced to strike the lad dead with a thunderbolt to save Earth.

At the time, of course, no one had any idea of the actual distance to the Sun, or how large the Sun might be, or whether you could carry it in a chariot. But there were those who would give their lives to find out.

"Willingly would I burn to death like Phaeton," wrote the Greek astronomer and mathematician Eudoxus, "were this the price for reaching the Sun and learning its shape, its size, and its distance."[1]

More than two thousand years would pass before astronomers were in a position to find out. By the year 1716, the relative distance of the planets from the Sun was known. For example, it was known that Venus is about three-quarters as far from the Sun as Earth is. But the actual distance from any planet to the Sun was not known. While the Sun and planets had been observed for thousands of years in their stately procession across the sky, the scale of the solar system was still a mystery.

In that year, the English astronomer Edmond Halley proposed a bold plan to finally find out. His plan was to use a method called parallax, which is still in use today. You can see for yourself how parallax works by trying a simple experiment.

Hold your arm outstretched, with your thumb sticking up. Look at your thumb with one eye open and the other eye closed. Now do the same with only the other eye open. Blink back and forth between your two eyes. Do you see your thumb appear to change position in relation to the more distant objects in the background?

Now bring your thumb closer to your face. Try blinking again, with first one eye open and then only the other eye open. Notice that your thumb jumps even more. Here's the principle: When you look at an object from two different vantage points, it appears to shift position in relation to the distant background. And the closer the object is to you, the greater the apparent shift. It is possible to figure out the distance between your thumb and face, provided you know the distance between your eyes and can measure how much your thumb appears to jump.

Halley's plan was to observe a celestial event known as the "transit of Venus" from widely separated vantage points. A transit is a type of eclipse in which the planet orbits in front of the Sun, as seen from Earth. The whole affair lasts only a few hours. Halley's plan was a little more complicated than the experiment you just tried, because, unlike your thumb, the planet Venus is always moving. So the timing of the observation was also critical. The plan involved noting the exact time that Venus began to transit in front of the Sun and the time that it finished its transit (Figure 1.1). By observing the transit from two different vantage points on Earth, it would be possible to use the parallax method to determine the distance to Venus—and, by extension, to the Sun and the planets. It would at long last be possible to determine the scale of the solar system.

There is a catch, however. As seen from Earth, the transit of Venus occurs only twice a century. There is a pair of transits eight years apart, and then nothing for another century.[2] The aging Halley realized that he would not live to make the observations himself, since the next pair of transits commenced in 1761.

"I recommend it, therefore," he wrote, "again and again, to those curious Astronomers, who (when I am dead) will have an

Figure 1.1. The Sun, viewed as the planet Venus (black disk) passes in front of it. By carefully observing this event from different vantage points on Earth, one can determine the distance to Venus and to the Sun.
Credit: NASA / Solar Dynamics Observatory.

opportunity of observing these things, that they would remember this my admonition, and diligently apply themselves with all their might to the making of this observation . . . and then, that having ascertained with more exactness the magnitudes of the planetary orbits, it may redound to their eternal fame and glory."

Halley urged that many observations be made at different locations, lest a single observer "be deprived, by the intervention of clouds, of a sight, which I know not whether any man living in this or the next age will ever see again; and on which depends the certain and adequate solution of a problem the most noble."[3] By getting measurements from as far apart as possible, the combined results would be more accurate.

For the transits of 1761 and 1769, many astronomers traveled the globe to make measurements, but I want to mention just three expeditions. Each team traveled to a destination that would be considered a vacation spot today. One team went to San José del Cabo, in the south of Baja California in Mexico; another went to Pondicherry, India; and a third traveled to Tahiti, in the South Seas.

It sounds glorious, but in the eighteenth century, a sea voyage was often a death sentence. Among the dangers were shipwrecks, storms, pirates, warships, diseases, and unfriendly native people — not to mention the rigors of life aboard a vessel for many months at a time.

The French astronomer Jean Chappe d'Auteroche was sent by the Académie des Sciences to San José del Cabo. He had hoped for a South Seas post, but that was not to be. After a strenuous eight-month journey, he and his team arrived at San José only to find the town seized by a deadly epidemic, probably typhus; many of the villagers had already died. He faced a life-and-death decision: stay and risk the contagion, or find another suitable location and risk not setting up his observatory in time for the transit.

What would you have done?

The intrepid d'Auteroche decided to stay. Yes, he did fall ill, but he was able to rise from his sickbed on the day of the Venus transit to make his observations and measurements. "I know that I have only a little time left to live," he wrote in his notebook, "but I have fulfilled my aim and I die content."[4] And, indeed, d'Auteroche, just forty-one years old, died, along with two Spanish naval astronomers and three-quarters of the Spanish and Native American villagers.

Among the two members of his team to survive was the artist Alexandre-Jean Noel, who sketched a grim portrait of the funeral procession. In those days of exploration, long before photography was invented, artists were as essential to an expedition as were the scientists, because they helped document the plants, animals, and topography of strange new lands.

The second team, sent to India, fared marginally better. The French astronomer Jean Baptiste Le Gentil had planned a career in the church, but after hearing an inspiring lecture about astronomy, he became a researcher instead. In 1760, not yet forty years old, he left family and fiancée and boarded a French warship headed for Pondicherry. After months at sea, and after braving furious storms, Le Gentil learned that the British had blockaded and finally captured the port of Pondicherry. Not able to make it to shore, he was forced to observe the transit of Venus from the deck of his rolling ship, from which he could not make any useful measurements.

Incredibly, Le Gentil decided to stay in the area for another eight years, in order to observe the next transit of Venus in 1769. After all, he had braved storms and warships on his long voyage to India, so he was determined to see his noble mission through.

Long years passed. Finally, the second transit of Venus was at hand. In Pondicherry, the months leading up to the transit had been blessed with beautiful weather. At two in the morning on the day of the long-awaited transit, Le Gentil awakened and arose. "I saw with the greatest astonishment," he wrote, "that the sky was covered everywhere. . . . I had exiled myself from my motherland, only to be the spectator of a fatal cloud, which arrived in front of the Sun at the precise moment of my observation, snatching from me the fruit of my efforts and exertions."[5]

When Le Gentil returned to France, eleven years after he had left, he was greeted with even worse news. His family, thinking that he had perished on the voyage, had divided up his estate. His fiancée had moved on. Two centuries later, his travails were the subject of *Transit of Venus,* a play by Maureen Hunter, and an opera of the same name.

The British astronomer Charles Green had a considerably higher-profile adventure in his quest to find the distance to the Sun. In 1768, he was to voyage to Tahiti on the ship *Endeavour,* under the command of the young captain James Cook, who had never before undertaken such a long voyage. The expedition was financed in part by the Royal Navy and the Royal Society, with the explicit goal of observing the transit of Venus from the South Seas.

Also on the voyage was a young entrepreneur named Joseph Banks. The son of a wealthy landowner, Banks had inherited his father's fortune as a youth. Rather than spending his inheritance on the high life at home, he put it to use hiring tutors to teach him the sciences, especially botany. He outfitted the expedition with his own team of eight assistants and with £10,000 of his own money—about four times what the ship itself was worth. As the *Endeavour* set sail in August 1768, Banks wrote in his diary, "We took our leave of Europe for Heavens knows how long, perhaps Forever."[6]

Point Venus in Tahiti is now famous for the observations that were made there. Despite foul weather on the days leading up to the transit, Green was overjoyed that the day of the transit was as beautiful as could be, with brilliant sunshine and not a cloud in the sky. According to a team member, Green and Captain Cook made their measurements "without one single impediment, excepting the heat, which was intolerable; the thermometer which hung by the clock and was exposed to the sun as we were, was one time as high as 119°."[7]

Captain Cook's voyages of discovery would crown him as among the greatest explorers of all time. Centuries later, he would serve as the model for the science fiction hero Captain James Kirk of the starship USS *Enterprise*.

Banks, just twenty-five years old when he set sail, returned to even greater acclaim than Cook. His name became synonymous with botany. Some of the plants adorning your garden or windowsill—such as magnolias, hydrangeas, and fuchsias—are likely to be descendants of the specimens that Banks and his team brought back from their many expeditions. He became an adviser to King George III, established the world-renowned Kew Gardens, and served as head of the Royal Society for forty-one years.

Green, on the other hand, never saw home again. He died at age thirty-five, along with the two artists and nearly half the crew on the voyage, many perishing from dysentery and malaria after a stop in what is now Jakarta, Indonesia. Green's observations and scientific notes on the transit have been criticized as disorganized. But of the six hundred scientific papers that resulted from the transit of Venus, only five teams' observations were selected to make the final calculation for the distance to the Sun, and Green's was one of them. D'Autoroche's was another.

Did the lives that these young scientists risked and lost make any difference to posterity? Absolutely, for many reasons, but here are three. First, their findings revealed the grandeur and immensity of our solar system, our own backyard in the universe. The Sun, we learned, is not just a brilliant light in the sky; its monstrous volume could swallow one million Earths. The Sun is so distant—about ninety-three million miles from Earth—that a jetliner would take seventeen years to get there. And yet our Sun is merely the nearest of all stars.

Second, their findings served as a yardstick for taking the measure of the universe beyond our solar system. They made it possible to determine the distance to nearby stars, then the size of our Milky Way galaxy, and ultimately the size of the observable universe.

But above all, these explorers showed us what it means to *need to find out:* to ask of the universe, Who or what are you that surrounds us, that created us? They epitomized the spirit of scientific discovery; they showed the lengths to which we are willing to go as a species to indulge a curiosity that is anything but "idle." They are a statement to the universe that we matter and we care. These Europeans may have journeyed to lands that were already known, already inhabited—but they brought all of humanity to a new place of the mind.

From Solar System to Galaxy

On a clear, dark night in the countryside, it seems as though you can see all of creation, with millions of stars glittering in the night sky. Alas, it is just an illusion. From the darkest spot on Earth, you can only make out a few thousand individual stars with your naked eye. There are more lights than that on some expressways at rush hour, more fireflies in a Vermont forest on a warm July evening.

But if you are lucky, you will also have seen the Milky Way, the great band of light arching across the sky like a trail of milk. With the aid of a telescope, Galileo was able to show that the Milky Way is made up of individual stars; they are so distant, they appear to merge into a continuous band when seen with the naked eye. Today we call this huge city of stars our galaxy, from the Greek word for milk. Our Sun is merely the closest star in this swarm of hundreds of billions of stars. Most people on Earth have probably never seen the Milky Way galaxy, because the majority of the world now lives in and around cities, where the sky is never very dark.

Until a generation ago, the planets that orbit our star, the Sun, were the only known planets in the universe. One could only speculate whether any other stars in the night sky hosted their own planets.

Today we know the answer. On average, *every star in the sky is orbited by at least one planet.* There are billions upon billions of planets in our galaxy, many of them the size of Earth. Already several thousand of these new worlds have been investigated by astronomers. We'll explore their significance in Part 3.

What Lies beyond the Milky Way Galaxy?

For most of human history, the Milky Way was considered the entire universe. It was not known what might lie beyond. But over the last century our understanding of the universe took an unexpected turn. We now have a panoramic view of the heavens that exceeds our wildest expectations.

This new era began with an event that took place in Washington, DC, in April 1920. The event was a debate, hosted by the National Academy of Sciences, between astronomers from two of America's premier observatories. The astronomers were Heber D. Curtis, from the Lick Observatory in California, and Harlow Shapley, from California's Mount Wilson Observatory.

The event has come to be known as the Great Debate, but the whole affair had an inauspicious beginning. Shapley was desperate to play down the event, because he knew that Curtis was an accomplished public speaker, while Shapley, still in his thirties, had relatively little experience before public audiences. Shapley had his eye on a newly available position at Harvard College Observatory, so he couldn't afford to be bested by another astronomer. In vain, Shapley suggested the event be called a "discussion" rather than a debate. He tried to shorten the length of time allotted, in order to diminish the event's significance. And he tried to arrange for the two participants to address different topics. No such luck.[8]

Both talks were given under the title "The Scale of the Universe." The debate would include an attempt to establish the size of the Milky Way galaxy, which was then poorly known. And it would try to pin down the nature of the strange, fuzzy patches in the sky, called spiral nebulae, that had long puzzled astronomers. Were these nebulae simply swirls of dust and gas in our Milky Way? Or could they be galaxies of stars—called island universes at the time—lying far beyond our own Milky Way galaxy? The term reflected the prevailing idea that the universe was represented by the stars, while the vast sea of space in which they sit was of little interest; space was simply there and was assumed to extend forever.[9]

Even the National Academy of Sciences had misgivings about the debate. The academy's secretary, C. G. Abbot, worried about the topic, fearing that the public might "care so little about island universes, notwithstanding their vast extent, that unless the speakers took pains to make the subject very engaging the thing would fall flat." He had suggested instead topics such as the cause of the Ice Ages, "or some zoological or biological subject which might make an interesting debate."[10]

Some had suggested a discussion of Albert Einstein's newly minted model of gravity—called the general theory of relativity.

But the academy would have none of relativity, figuring that fewer than a half dozen members of that august body would be competent enough to understand even a few words of what the speakers were saying. "I pray to God," said Abbot, "that the progress of science will send relativity to some region of space beyond the fourth dimension, from whence it may never return to plague us."[11]

As we will see, relativity did return, in full force, but its amazing predictions would rest in part on the outcome of this little-known debate in Washington.

Shapley focused first on the size of the Milky Way galaxy.[12] Building on a ladder of distance scales, starting with the size of our solar system, he was able to estimate the immense size of the galaxy. The distance from Earth to the Sun was known to be ninety-three million miles, but the distance across the width of the Milky Way was so large that it could not be conveniently measured in miles at all. Instead of miles, he used the distance that light travels in an entire year; a light-year is about six trillion miles. His estimate for the width of the galaxy was a bit too large, but close enough. Today we know that our Milky Way galaxy is at least one hundred thousand light-years in diameter. That's about 600,000,000,000,000,000,000 miles!

Shapley realized that the light-year was not merely a measure of distance; it was also a measure of history. "It tells us not only how far away an object is," he wrote, "but also how long ago the light we examine was started on its way." Thus we see the Sun not where it is but where it was eight minutes ago. "You do not see faint stars of the Milky Way as they are now, but more probably as they were when the pyramids of Egypt were being built; and the ancient Egyptians saw them as they were at a time still more remote." Shapley marveled at "the enormous ages of stars" and "the vast extent of the universe in time as well as in space."[13]

The distances within our galaxy are so vast that something like a *Milky Way Nightly News* show could not exist. If something newsworthy happened on the other side of the galaxy, a speed-of-light broadcast would not reach us for another sixty thousand years or so. And the news that does arrive tonight would refer to something that happened thousands of years ago. Not good for the ratings! The galaxy requires a quieter, more leisurely concept of news than we are used to.

Within the vast expanse of the Milky Way, where does our solar system lie? At the time, in 1920, many astronomers believed that the Sun and its planets were pretty much at the center of the galaxy, since there appeared to be equal numbers of stars in all directions. But Shapley's careful observations of more-distant stars told a different story. Our own solar system is a few tens of thousands of light-years from the center of the Milky Way. "We have been misled," he said, "into thinking that we are in the midst of things," much as ancient man was misled "into believing that even his little planet was the center of the universe, and that his earthly gods created and judged the whole."[14]

Shapley is widely considered to have won this part of the debate. But he had little to say about the nature of the mysterious spiral fuzzy patches in the sky—the spiral nebulae—and whether they might be distant galaxies. And what he did say was wrong.

"I will leave the description and discussion of this debatable question to Professor Curtis. . . . I prefer to believe that they are not composed of stars at all, but are truly nebulous objects." A colleague of Shapley's shared the view. "For my own part, I am still on the fence on the question. I very greatly doubt the visibility of half-a-million or more 'island universes.' "[15]

Curtis, on the other hand, marshaled the evidence that the spiral nebulae are in fact galaxies of stars in their own right, scattered throughout space far beyond our own Milky Way. He noted, for example, that when light from these objects was passed through a prism, their spectrum was similar to typical stars in our own galaxy. Therefore the nebulae were probably composed of stars. He noted the occasional occurrence of bright "new stars" in the nebulae, similar to the new stars, or novae, that appear from time to time in our own galaxy. He pointed out that our own galaxy was thought to have a spiral shape, so it was reasonable that the spiral nebulae might also be galaxies. And, based on Curtis's rough estimates of the distance to these spirals, the objects would be about as wide as our own galaxy. He noted that the fuzzy spirals often had an obscuring, dark layer around their outer edges, presumably dust, again probably similar to the dust in our own galaxy.[16]

And finally, he noted a curious feature of the nebulae. They were moving fast. Very fast. Much faster than typical stars in our galaxy. At the time, he didn't know what this meant. But in Europe, Einstein and a handful of other physicists were already developing theories that would eventually explain these motions—and turn our ideas about the universe upside down.

Today we know that the universe is filled to the brim with galaxies. Try this: Hold a grain of sand up to the sky at arm's length. Along the line of sight of that grain of sand, in just that tiny patch of sky, there are more than ten thousand galaxies, each containing billions of stars. In all, there are at least one hundred billion galaxies in the observable universe—and possibly infinitely more beyond what we do see.

To get a sense of the vastness of the universe, consider the size of just a single galaxy. The Andromeda galaxy is the nearest large galaxy to our own, and also the farthest object you can see with your naked eye (Figure 1.2). The whole galaxy spreads across a patch of sky about seven times wider than the full Moon. It would be the most dramatic object by far in the night sky, but it is so faint that we can only see the very center of the galaxy with the naked eye; it appears as a fuzzy patch of light.

The Andromeda galaxy is about twice the width of our Milky Way galaxy. At the scale of the image of Andromeda in Figure 1.2, a typical star in the galaxy would be smaller than an atom.

Thus, if you could image a galaxy precisely, without any light spilling over into adjacent pixels, the galaxy would seem to disappear. It is, after all, almost completely empty space—and that is exactly what you would see: nothing. Even if the galaxy were to move sideways at the speed of light, it would not travel far enough in our lifetimes for us to see it budge even one pixel's worth. Thus galaxies appear to be frozen in time. We can see evidence of past collisions and explosions and catastrophes, but we cannot directly see the galaxy as a whole moving *now,* as we observe in real time.

In retrospect, knowing what we know now, the Great Debate of 1920 was revolutionary. But at the time, it made little if any impression on the press. Perhaps there were too many other upheavals in 1920. The Great War, the war to end all wars, left anxiety in its aftermath. Sigmund Freud's psychoanalysis had become the rage. In the November presidential election, women would wield the vote for the first time in the United States.

Or perhaps the public simply cared little about island universes, as the academy's secretary had feared. Shapley himself was said to have little memory of the debate later in life. And Curtis's obituary

Figure 1.2. The Andromeda galaxy, the nearest large galaxy to our own Milky Way. *Top:* An image taken in ultraviolet light, showing the hottest stars and the regions where stars are actively being born. Credit: NASA/JPL–Caltech. *Bottom:* A close-up view showing a few of the roughly one trillion stars in the Andromeda galaxy. If the image were perfectly sharp, each star would appear smaller than an atom! The stars look close together, but they are trillions of miles apart. Credit: NASA, ESA, J. Dalcanton, B. F. Williams, and L. C. Johnson (University of Washington), the PHAT team, and R. Gendler (CC BY 4.0).

apparently didn't mention the event at all. But within the decade, a true revolution in our view of the cosmos would be in full swing.

Are We "Significant"?

When Shapley wrote up a summary of his presentation for publication, he noted that humans have been progressively displaced from our privileged position in the universe. First we were banished from the center of the solar system by Nicolaus Copernicus. Later we lost our position at the center of the Milky Way galaxy. Shapley wrote that as the system of stars has appeared ever larger, "the significance of man and the earth in the [heavenly] scheme has dwindled."[17]

It is worth digressing for just a moment to consider whether this is true. Has our significance "dwindled"?

Imagine for a moment that you had never seen the night sky. Let's say that every day was a beautiful, sunny day, but that by dusk the mists rolled in and the rains came, so that neither you nor anyone else had ever seen the stars.

Would you be different in any way?

I have asked this question of many people over the years, and I am always surprised at the responses. At first, people tend to reach into their bag of facts and draw one out. Without stars, someone will mention, it would have been difficult to navigate, so the New World might still be a question mark on European maps. Our calendar would certainly be different, since five days of the week are named after the planets. And it would have been harder to keep track of the seasons and to know when to plant crops.

But when I repeat the question—would *you* be different emotionally?—the answer is very different. "I love looking at the night sky," a person will typically say, "because it makes me feel so small, so *insignificant*."

And that is the word that keeps coming up again and again. *Insignificant*. There is an innocent interpretation to the word which we all can understand and share, such as, "My everyday cares and aspirations are insignificant in the grand scheme of things." There is almost a thrill in being infinitesimally small in comparison to the grand sweep of the universe.

But there is also an edgier side to *insignificant,* the one that connotes lack of meaning or import.

Here is an alternative view of our significance. Yes, we feel tiny when we see the night sky. But, remarkably, our collective gaze fills the entire observable universe. Thanks to our telescopes and other instruments, we can now observe all the way out in space and all the way back in time, essentially to the moment of the Big Bang. *We see virtually everything that the laws of nature allow us to see.*

It did not have to be that way. Earth could have been perpetually cloudy, as many planets are. The solar system could have been dusty; the galaxy could have obscured our view beyond. The universe itself could have been shrouded and opaque. But nature has arranged that, in this immense universe, at this particular time in cosmic history, each of us, no matter how tiny we are, can roam the entire universe with our gaze and our minds.

2

Galaxies Misbehave

The universe is so vast, so old, and so silent that it might seem impossible to learn anything about where it came from. How could the billions of galaxies that lie so far beyond our Milky Way galaxy be intertwined in the slightest with our own cosmic origins? Remarkably, nature has provided us with the clues we need to decipher our origins—and it even allows us to peer back in time directly with our telescopes. It turns out that every particle, every atom, every living thing in the universe is bound to us through a common origin.

The first clue to our cosmic origin lies in a mundane question: How are the galaxies moving through space? By knowing how galaxies move today, it might be possible to deduce what they were doing in the past. Nature is full of surprises, so it helps to first try to predict how galaxies might be moving.

Figure 2.1 shows three plausible possibilities for how galaxies might be moving, along with the reasoning in favor of each. Take a moment to consider each of these possibilities. Which one appeals

1. Galaxies might all be moving toward us

"Gravity attracts, so I think that everything should be falling together. The galaxies out there should be moving toward each other, and toward our own Milky Way galaxy. Sooner or later, the galaxies will crash together!"

2. Galaxies might not move at all

"I think the universe must be in perfect balance. If the galaxies were falling toward each other, wouldn't the universe have collapsed long ago? Balance rules! I think the galaxies are not moving at all."

3. Galaxies might move randomly

"I think the galaxies are too far apart for gravity to make a difference. Maybe the galaxies are all moving randomly, like a swarm of bees on a hot day. Some galaxies move toward us, others move away."

Figure 2.1. Three reasonable predictions for how galaxies might be moving through space, in relation to our own Milky Way galaxy. The schematic illustration shows four galaxies, with the Milky Way galaxy in the center.
Credit: Composite images adapted from Clipart Library.

most to you? Feel free to come up with a different prediction of your own, but justify it with a reason.

Have you committed to a prediction? Are you comfortable with the reason? All of the possibilities seem plausible, so it may be difficult to choose among them. Spoiler alert: we're about to find out which possibility nature actually chooses.

In 1913, an Indiana-born astronomer named Vesto Slipher, working at the Lowell Observatory in Arizona, measured the speed of the Andromeda galaxy, our neighboring galaxy.[1] "We may conclude," he wrote, "that the Andromeda Nebula is approaching the solar system with a velocity of about 300 kilometers [about 200 miles] per second."[2]

This result was from just one galaxy, of course, but it seems to support Possibility 1 (gravity rules). It certainly makes sense that the Milky Way and its nearest large neighbor would be falling toward each other: What else would gravity do, if not pull them together? After the next few galaxies were measured, however, it was found that some galaxies are moving toward us, while others are moving away. So if you chose Possibility 3 —that galaxies move randomly— take heart.

But not for long. As more-distant galaxies were probed, the pattern began to change again. The great majority of galaxies were seen to be moving *away* from us. In fact, today we observe that— with the exception of the handful of nearest galaxies—*every galaxy in the universe is moving away from us.*

This bizarre finding, an "expanding universe," is not on our list of possibilities. Why should it be? No scientist in the world, not even Albert Einstein, would have predicted that the galaxies are rushing apart from each other—and I assume that you didn't predict it either. After all, gravity should attract, not repel. And if gravity attracts, why should the galaxies all be flying farther apart? What in the world is going on?

By 1929, an even stranger finding emerged. Working at the Mount Wilson Observatory in California, the astronomer Edwin Hubble and his assistant, Milton Humason, discovered a pattern to the expansion: the distance to each galaxy is proportional to its speed away from us.[3] As a consequence, all the galaxies would have originated from the same point in space billions of years ago, based on estimates of the galaxies' current speeds. (This is similar to a road race in which the runners start in the same place, but as time goes on, the faster runners are proportionally farther from the starting line.)

What could possibly be causing the expansion of the universe? At first glance, it might seem as though there had been a cosmic explosion in the deep past, hurling matter outward like the shards of a bomb or hand grenade. The clumps of matter that were hurled the fastest would naturally be the farthest away from the origin of the explosion. Slower-moving clumps would be closer to the site of that catastrophe.

And, in fact, many scientists have used the explosion metaphor in popular accounts. "The galaxies are not rushing apart because of some mysterious force," wrote the physicist Steven Weinberg. "Rather, they are moving apart because they were thrown apart by some sort of explosion in the past. . . . The expansion is just the effect of the velocities left over from a past explosion."[4]

But there is something very troubling about this picture of an explosion. If it were true, then what we see when we look out at the universe would depend on what shard of matter we happened to be on (or in). If we were in the fastest-moving galaxy, here is what we would see: When we looked in one direction, we would see the empty space into which we were hurtling. Looking in the opposite direction, we would see all the slowpoke shards behind us. *But that is not what we observe.* The universe looks equally populated with galaxies, no matter which direction we look.

Perhaps we happen to be at the center of the explosion and have not moved at all, so we see all the shards moving away from us symmetrically. But why would our galaxy happen to remain at the center of the explosion? That doesn't make sense.

Hubble found it hard to believe his own findings. "It is difficult to believe that the velocities are real; that all matter is actually scattering away from our region of space," he wrote in July 1929. "It is easier to suppose that [our observations could be better explained] by some property of space or by forces acting on the light during its long journey to the Earth. The problem is now in the hands of the theorists."[5]

And there's another problem. We would expect the flash of light from an explosion to have sped off into space and be long gone, since light outpaces any form of matter. But, as it turns out, that's not the case: We do indeed observe faint light of just the sort we would expect from an explosion. But the astounding thing about this ancient light is that it comes to us continuously from *all* directions in space.

Consider how paradoxical this is. We observe galaxies flying apart from each other, in a pattern that indicates there was a cosmic explosion long ago. Yet no matter which direction we look, we see the same general landscape of galaxies, and the same pattern of light. If this is confusing to you, don't worry—you're in good company. More people have been confused by this than by any other aspect of cosmology. And with good reason: it is not possible to reconcile Hubble's observations, or to visualize the expansion of the universe, simply by rearranging what you already know.

We need a completely new idea.

It is a remarkable coincidence that, at the same time these observations were throwing the scientific world into disarray, new ideas were being developed independently that would explain the observations and drastically change our view of the universe. It turns out that the galaxies are not moving through space. It is space itself that is coming into existence, welling up between the galaxies. We are witnessing evidence for a universe in creation. The creation of the universe was not an event that only happened long ago and far away. It is happening now, as you read this.

In the next few chapters, we will trace the origins of the new idea that we'll need in order to make sense of Hubble's observations. It is a simple idea, and it will help us resolve the paradox presented here. But, much more importantly, it will tell us something extraordinary about our universe. So we will put aside our observations about the universe, take a deep breath, and dive into the world of ideas.

3

What's the Big Idea?

You might think that revolutionary ideas about space and time began with Albert Einstein, but as we'll see, even the most revolutionary ideas can have a long and innocent gestation. To fully understand the concept we'll need, let's pick up the story in 1623, as one of William Shakespeare's most charming plays makes its debut.

The play is *As You Like It*. Rosalind is strolling through the forest with her friend Orlando. She notes that if there were lovers in the forest, their "sighing every minute and groaning every hour" would mark out "the lazy foot of Time as well as a clock." Her friend wants to know why time should flow lazily. "Why not the *swift* foot of Time?" he asks.[1]

Rosalind's reply is three hundred years ahead of her time: "Time travels in divers paces with divers persons." She explains that, for a young woman about to be married, seven nights seems to pass as slowly as seven years. Yet, for "a thief to the gallows," every footfall is too fast. And, for a lawyer asleep in the summer when no cases are pending, time seems not to pass at all.

Of course, Shakespeare is referring to the psychological perception of time, not physical reality. Although we have all experienced what he describes, we would still surely say that, regardless of what our mind tells us, there is but one, universal scale of time, one march of the clock that carries us from cradle to grave. Nevertheless, there was a new thought in the air, the daring thought of time flowing faster or slower for different observers. Who is to say whether time flows swiftly or with a lazy foot?

In less than a century, the physical nature of space and time would become the focus of a historic and bitter debate that holds the key to the idea we'll need. The whole episode came to pass through the efforts of one of history's most remarkable women: Caroline of Ansbach. If you have ever daydreamed about having dinner with a figure from the past, she should be on your short list.

Caroline was born to the ruler of a small German principality. Although she was orphaned as a child, she was brought up in the court of King Frederick I of Prussia and studied with the great German mathematician and physicist Gottfried Leibniz, with whom she later corresponded for years.[2] She moved to Britain in 1714 after her husband, George Augustus, became the prince of Wales. He would go on to become King George II and she, the queen of England and grandmother of King George III. (The rest, as they say, is history—American history.)

Having a keen mind and great curiosity, she was familiar with Isaac Newton's work, as well as with Leibniz's. She realized that Leibniz and Newton held incompatible views about the nature of space and time. To put it bluntly, one or the other physicist must be wrong. She attempted to resolve their dispute by arranging a debate in the form of an exchange of letters, starting in 1715.[3]

Some royalty would have settled a debate by fiat, or executive order, or an appeal to one religion or another, or by even by force.

(Louis XIV was said to have had his most fearsome cannons inscribed with the motto "The final argument of kings.") But Caroline didn't want the mere semblance of being right. She wanted to know who was in fact right.

Being diplomatic, Caroline decided to give Leibniz's first letter to a disciple of Newton's, rather than to the great man himself. Leibniz and Newton were bitter rivals—each having claimed credit for inventing calculus—and their sparring included a great deal of speculation about God's relationship to space. So she gave the first letter, the opening salvo in their debate, to a theologian and scientist named Samuel Clarke, who had translated Newton's *Opticks* into Latin. Clarke became Newton's proxy in the debate.

Amid all the philosophical, theological, and rhetorical discussions that ensued, there bubbled up a single idea that would eventually change the way we see space and time. It was Leibniz's idea: "These gentlemen," he wrote, referring to Newton and Clarke, "maintain that space is a real absolute being. For my part . . . *I hold space to be something merely relative, as time is*" (emphasis added).[4]

What in the world did Leibniz mean? To understand his idea, and have some fun at the same time, try the puzzle in Figure 3.1. The photograph shows a vintage car from the 1950s. Can you tell whether the car in the photo is real—or just a scale-model toy that you might find in the driveway of a dollhouse?

Not so easy to tell, is it? In fact, merely by looking at the photo, it is not possible to know the car's size for certain. *To tell, you would need to compare it to something else in the real world.*

If there were a yardstick in the scene, you might think you could easily determine the car's length. But, on second thought, you would have no way to tell whether the yardstick was just a scale model as well.

Figure 3.1. Is this a real car, or a scale model car? How about our universe? At what scale is it constructed, and how could you ever tell?
Credit: CC0 1.0.

Here's a more puzzling question: At what scale are you and I and the rest of the universe constructed? Are we large-scale people? Or dollhouse-size people? If we have nothing to compare ourselves to, size seems to have no meaning. Large or small compared to what?

A final question: Is the universe constructed at the same scale everywhere? Could the scale of length—for example, the length of a ruler—vary from place to place? Could one location be normal scale while another was dollhouse scale?

Could the scale of the universe change with time, as well?

To Newton, the answer to these questions would be obvious. Space is "absolute," as he put it: There is a God-given scale to the universe; a yardstick is the same length no matter where it happens to be located. It is the same length on Earth as it is on the Moon. It doesn't suddenly shrink or grow when you move it to another location. If you ask, "Why should a yardstick have the length it does and not be longer or shorter?" Newton would answer, Because that is the way that God created a yard.

Newton likely believed this for two reasons. First, it seems obvious. You don't appear to shrink when you walk across the room or sail to Australia. Second, the physics that Newton had carefully developed *required* a constant, reliable backdrop of space, for which the scale of distance was everywhere the same. He wasn't about to give up the bedrock foundation of his work in physics.

But Leibniz was adamant. The issue is not what you imagine God wants, he reasoned; it is what human beings experience and can measure. And the only way to tell whether your yardstick, you, and everything else shrinks to dollhouse proportions when you are on the Moon is to compare it with something else. "I hold space to be merely relative, as time is."

If Caroline weighed in on the debate, we have no record of it. In any event, Leibniz died in 1716, shortly after his last letter was sent. His funeral was sparsely attended, and neither the Royal Society nor the Berlin Academy of Sciences, of which he was a member, honored his life.

Newton's view of space and time held sway for the next three hundred years. But Leibniz would turn out to be right. His germ of an idea—that space and time are not fixed but instead might be malleable and stretchable—was in the air again, would not go away, and would eventually change the way we see the universe. Let's see how The Idea evolved.

Fast-forward a century, to an auditorium at the University of Göttingen in Germany. The Idea has taken root in the fertile mind of a shy twenty-seven-year-old named Bernhard Riemann. He had originally studied theology, in order to please his father, a

Lutheran minister, but then changed his mind and asked for permission to study mathematics instead. After finishing his studies, it only remained for him to give a lecture on mathematics to earn the degree he needed for a post as a lecturer at Göttingen. So on June 10, 1854, Riemann delivered a lecture called "On the Hypotheses Which Lie at the Foundations of Geometry." It was said that few in the lecture hall realized the importance of the work, save for Riemann's mentor, Carl Friedrich Gauss. But soon, the work would be the cornerstone of a revolution.

As part of his lecture, Riemann had the audacity to question Newton's and Euclid's assumptions that lines have a length independent of position. He developed the geometry of possible worlds—worlds of the mind—in which the length of an object could actually change with location. In such a world, a car in one location might become a toy-size car if you moved it to another location.

To visualize what Riemann had in mind, you need only recall the Mercator map of the world that you saw in your schoolroom as a child. On that map, the scale of distance depends on latitude. Up in Greenland, a line representing a thousand miles is much longer than a thousand-mile line at the equator. The geometry of this flat map is not the Euclidean geometry we learned in school. (For example, on this map, the shortest distance between two points is not a straight line; it is one that looks curved to the eye.) The map is an artificial world in which scale varies with location.

Of course, we all know that this varying scale of distance is an artifact; it comes from stretching out the spherical surface of Earth onto a flat surface. In the real world, the length of a line really does seem the same everywhere, just as Euclid assumed two thousand years ago, and as Newton took as a God-given fact.

But Riemann looked beyond the everyday world through which we move. He wondered whether the world of the very large or very small might obey the very different geometry he had just developed.

Riemann knew that only observation and experimentation could determine how nature is constructed. Sweeping away two thousand years of Euclidean geometry, he said to his audience, "We are quite at liberty to suppose that [the ways that length is measured] in the infinitely small do not conform to the hypotheses of geometry . . . and *we ought in fact to suppose it, if we can thereby obtain a simpler explanation of phenomena*" (emphasis added).[5]

Riemann was wonderfully diplomatic in his approach. He was not denying what Newton assumed to be true; he was merely starting from where the great man had left off. "The answer to these questions," he wrote, "can only be got by starting from the conception of phenomena which has hitherto been justified by experience, and which Newton assumed as a foundation, and by making in this conception the successive changes required by facts which it cannot explain."[6] That is, when you need it, my new view of geometry is there for your use.

Cleverly, Riemann ended with a flourish designed to outflank his critics. His work was merely useful, he claimed, in preventing the work started by Euclid and Newton "from being hampered by too narrow views, and progress in knowledge of the interdependence of things from being checked by traditional prejudices."[7] Who would want to be narrow-minded and prejudiced?

Unfortunately, there was not a shred of evidence to support Riemann's geometry—nothing to suggest that the world was anything other than what Euclid and Newton supposed: The length of a line, the length of a ruler, the height of a person obviously don't change just because you move them to a new location. His work was merely a mathematical curiosity.

Or was it?

Within a decade, this great mathematician would succumb to tuberculosis, leaving behind his wife and small children. But his work, the mathematical foundation that he gave to The Idea, would live on in the next century and take root in the fertile mind of Albert Einstein.

4

Einstein, Gravity, and the Universe

The early pioneers had merely argued that an object's size *might* change from location to location. But starting in the late nineteenth century, a series of discoveries led to the breathtaking conclusion that the size of every object *must* depend on its location, even on its speed. What was once merely a wild idea was about to become a revolution.

The big breakthrough was a discovery about the most common thing in the world: light. It was established that light has a paradoxical property shared by no other object: it travels through space with a constant speed that appears the same to all observers, regardless of how fast or slowly they are moving.

To see why this so strange, consider your own speed right now. If you happen to be sitting down, you might think your speed is zero. But now you remember that you are sitting on an airplane, so a person on the ground would observe you moving at six hundred miles per hour. And an astronaut arriving from Mars would see you speeding around the Sun at more than sixty thousand miles per hour. Clearly, your speed depends on who is doing the observing.

Yet the speed of light is utterly different. It does not depend on who is measuring it; it is the same for all observers. No matter how much you try to catch up with a particle of light, it still races away from you at the speed of light, as though you were standing still. How is that possible?

The paradox was solved by Albert Einstein and the Dutch physicist Hendrik Lorentz; their work culminated in 1905 in what became known as Einstein's special theory of relativity. (The word *relativity* is unfortunate and misleading, because the theory is founded on the observation that light's speed is not relative to anything; it is a constant of nature.)[1]

The theory required giving up Isaac Newton's long-held belief that there is a single, universal scale of length and time. Instead, it must be that *two observers who are in relative motion do not share the same scale of length and time.* It was the first vindication of Gottfried Leibniz's claim that length only has meaning in comparison between two objects. No one would have imagined that speed has anything to do with it. But that is just the way nature works.

In our everyday world, we don't notice these changes in the least, because we move at a snail's pace compared to light, so the effect is so tiny that we don't see an object shrink when it moves, and we certainly don't see time slow down. But, for very fast moving objects, these effects become significant. And light itself travels so fast that its flow of time stops completely: A particle of light has no past and no future; it is eternally in the present. It cannot evolve.

It is a remarkable fact that everything strange that we are about to learn about the universe, all of the rearrangements of our thinking that must take place, follows inexorably from the strange behavior of light.

How bold is the human imagination to imagine that distance and time are not what we observe every day! Bernhard Riemann had speculated that we might have to go to the world of the infinitely large or small to see any strangeness, but here it was the world of the very fast in which space and time behave so strangely.

To this day, Einstein's discovery about space and time is seen as an almost hallowed achievement because it marked the end of innocence about the universe, and the realization that the universe is much stranger than we thought.

Years ago, when I was a youth and tried hiking across Switzerland, I rested wearily against a storefront in the lovely town of Bern, I felt something cold against my back. It was an old bronze plaque that read, "In this house, in 1905, Albert Einstein developed his theory of relativity." As a young scientist, I had the feeling of suddenly being at the center of the cosmos, the thrill that something tremendous happened here, and not just for Einstein but for all of us.

Einstein's 1905 discovery was merely the opening act. Physics is a web of connected concepts, so when one cherished idea falls, others topple as well, one after the other, like a row of dominoes. For example, Einstein quickly realized that Newton's model of gravity was inconsistent with the observation that the speed of light is the same for all observers. Newton's model had successfully described the fall of a proverbial apple, the arc of a cannonball, and the lumbering orbit of a planet; it had stood unchallenged for two hundred years, but now it would have to be modified.

Einstein's theory of relativity was like a monkey wrench that had been thrown into the elegant works of physics. To repair the damage, and to restore the elegance of physics, he set out to construct a new model of gravity—one that was compatible with the strange behavior of light and with all the consequences that followed from it. It was a monumental undertaking, a slog across a minefield strewn with problems. Not only did space and time behave strangely but, as a consequence, so did energy and mass; Einstein had to work out an entirely new relationship between energy and mass, one that would preserve the cherished rule that energy can be neither created nor destroyed but merely changed from one form to another.

Einstein wrote that these problems "should not frighten us away from the further pursuit of the path we have taken."[2] For ten long years, he struggled to find a model of gravity that would unite space, time, and matter in one framework that preserved the conservation of energy and provided a self-consistent description of gravity.

Einstein's struggle was unique in the history of science. Previously, scientists had the benefit of hundreds of observations and experiments to guide their theories, their models of how the world works. Newton's laws of motion and gravity drew on centuries of observations about the motions of objects both earthly and celestial. James Clerk Maxwell had the benefit of a century of laboratory experiments to guide his theory of electromagnetism. Charles Darwin traveled the globe, meticulously observing and recording as he developed his theory of evolution.

But Einstein had no such recourse to either nature or the laboratory. The world around him seemed to support Newton's and Euclid's view: a foot is a foot, a mile is a mile, no matter where you happen to measure it. Instead, Einstein was guided by deep physical insight, by intuition, by aesthetics, and by a mere handful of insights about how the world ought to work. They were convictions raised to the status of principles: An accelerating frame of reference should be indistinguishable from a gravitational field. The laws of physics should have the same form regardless of the motion of the observer.

Furthermore, Einstein was unfamiliar with the mathematics that he needed. His friend Marcel Grossman helped him, and so did Michele Besso.[3] And above all was the mathematical foundation laid by Riemann, whose strange geometry was precisely what Einstein required. Einstein praised "the genius of Riemann, solitary and uncomprehended," and Riemann's "new conception of space, in which space was deprived of its rigidity, and in which its power to take part in physical events was recognized as possible."[4]

Finally, in the fall of 1915, Einstein completed his model of gravity. The very first thing you want to know about any new model in physics is how the quantities that it describes are related to each other. In this case the quantities are mass, length, and time. In Newton's conception of physics, there is no relation between them; they are completely independent of each other. Whether you weigh a hundred pounds or a million pounds makes no difference to the space and time around you, according to Newton. They are not affected in the slightest.

But Einstein discovered that mass, length, and time are *not* independent of each other. They are joined at the hip by a precise and profound relationship. Mass, it turns out, distorts nearby lengths and the flow of time. The greater the mass, the greater the distortion. (The model is a little more complicated than this: a spinning mass not only stretches nearby objects but twists them as well; fortunately we won't need these details.)

The second thing you want to know about any new model in physics is what the model predicts. Does it accurately describe the real world? To test the model, you have to apply it to a specific example and compare its prediction with what you actually observe.

Einstein's model is conceptually simple, but it is embodied in a set of equations that are very difficult to solve for any specific example. At the time, very few people even understood Einstein's model, let alone were able to apply it. Later on, Einstein's model of gravity would be called the most creative achievement in the history of science, but at the time, it was an inscrutable curiosity.

That makes it all the more astonishing that the first application, the first test, of Einstein's model of gravity came not from the great man himself but from a most unlikely person in the most appalling conditions. The man was Karl Schwarzschild, a German physicist and mathematician who found himself at the Russian front in World War I in the autumn of 1915. Schwarzschild had been pressed into service calculating artillery trajectories. While surrounded by the carnage of war, Schwarzschild had somehow managed to master Einstein's model of gravity in his spare time and even to apply it to solve two outstanding problems. What astounding focus the man must have had!

The first problem was to calculate the orbit of the planet Mercury. Newton's physics had long explained the orbits of all but one of the known planets, the exception being Mercury, the planet closest to the Sun. Mercury's orbit showed a tiny discrepancy— less than 1 percent, from what Newton predicted, but a discrepancy nevertheless. Presumably this oddity was caused by the mass of the Sun, which distorts the space and time through which Mercury passes.

Could Einstein's new model of gravity account for the discrepancy? Einstein had started the calculation, but it was Schwarzschild who derived a complete solution.

A few days before Christmas in 1915, Schwarzschild sent a letter to Einstein from the battlefield. "It is a wonderful thing," he wrote, "that the explanation for the Mercury anomaly emerges so convincingly from such an abstract idea."[5] It was the first indication that Einstein's abstract model had the power to describe the real world.

Schwarzschild quickly went on to a second problem. He calculated the distortion of space and time around a spherical mass, such as Earth. His results showed that Leibniz and Riemann were on the right track: The universe is not built at a single scale. The length of a line depends on its location! Schwarzschild's results predicted that everything—a yardstick, an atom, a person—shrinks in scale the closer it is to the mass of Earth.[6]

This distortion is tiny. A person at sea level is shorter than a person atop Mount Everest by less than the size of an atom![7] And your watch is slower at sea level by less than a trillionth of a second. This tiny distortion of time has actually been measured using sophisticated atomic clocks.[8]

In his battlefield letter to Einstein, Schwarzschild concluded with a few words that speak volumes about the triumph of the human soul in the face of evil. "As you see, the war is kindly disposed toward me," he wrote, meaning that he was still alive, "allowing me, despite fierce gunfire . . . *to take this walk into your land of ideas*" (emphasis added).[9]

By spring, Schwarzschild would be dead. He was just forty-three years of age. The land of ideas through which he wandered, however, is vast and eternal. Schwarzschild's work would form the basis of our understanding of nature's most bizarre phenomenon, the black hole. And it would inform modern technologies such as the Global Positioning System (GPS), which takes into account even the tiny distortion of space and time surrounding Earth. If GPS did not take this distortion into account, it would guide you to your neighbor's driveway instead of your own!

The distortion of space can actually be detected directly. In 1916, Einstein predicted that massive objects can create gravitational waves that travel through space—much as a rock dropped in a pond creates waves that travel through water. Gravitational waves are often described as ripples in the fabric of space-time, but they can be visualized as a traveling distortion in the scale of distance. They are passing through you right now, alternately shrinking and stretching you by an imperceptible amount.

Until recently, it was thought that gravitational waves might be too faint to detect. But in 2016, after decades of effort, a large team of researchers announced the first detection of gravitational waves, coming from the thunderous merger of two black holes more than a billion light-years away.[10] The remarkable facility built to detect these waves is called the Laser Interferometer Gravitational-Wave Observatory (LIGO). One of the LIGO detectors is positioned in Washington State, and the other detector is in Louisiana. Each consists of a pair of arms several miles long that serve as rulers. As a wave passes by, the arms change length by *a million times less than the width of an atom*. Incredibly, this tiny distortion has now been measured and analyzed, providing information about a cosmic collision that took place in the far reaches of space long before life on Earth had crept from sea to land.

If we were sitting around a campfire, swapping stories about the universe, this would be the moment at which someone would ask, "What's the point of nature's story? Why should nature tweak space and time just a little bit? Surely not to make Mercury's orbit wobble a bit, or to make our GPS a bit harder to develop?"

Nature doesn't need reasons for what it does, of course, but if you ask *why,* nature sometimes provides unexpected and exhilarating answers. That is the case here: If nature didn't tweak space and time this way, you and I wouldn't be here. In fact, the universe as we know it would not exist. That tweak—that strange idea that mass distorts the scale of space—will turn out to be the way the universe creates more of itself.

Let's see how that works.

Einstein's Model of Gravity, Applied to the Universe

The hallmark of a good scientific model is that it accounts for what we actually observe in nature. But the hallmark of a truly great model lies in the unexpected predictions it makes—what it tells us about nature that we didn't know, might never have guessed, or might not have believed.

By this measure, Einstein's model of gravity stands alone. For it predicted the existence of three new phenomena in nature that are so outrageously strange that even Einstein didn't believe they actually existed, and neither did anyone else at the time. These three wonders of nature are black holes, dark energy, and the one we turn to now: the expanding universe.

In 1916, Einstein applied his new model of gravity to the universe as a whole. It's not that he was feeling grandiose that day. It was because the dynamics of the universe was one of the few problems he could solve using his model of gravity. The problem of the universe turns out to be manageable if you make the right simplifications.

Here is the problem: Idealize the universe as being uniformly filled with matter and energy. What happens to space and time in such a universe? Specifically, what happens to the *scale of length* as time passes? Does the universe remain at the same scale? Or does the scale change with time?

You will have noticed that the universe is *not* uniformly filled with matter. There are clumps of matter here and there—a star, a planet, you and me—and then vast stretches of space between the stars and galaxies. But if you consider a big enough chunk of space—say, a hundred million light-years on a side—then there is pretty much the same amount of matter and energy in that big chunk no matter where the chunk comes from. So, from a big-picture point of view, the universe is uniform, just as the air in a room is uniform if you don't look too closely at the individual molecules. Einstein didn't know it at the time, but his simplified universe turns out to be a good approximation to the real thing. By making the simplifying assumption that mass is uniformly distributed throughout the universe, Einstein was able to solve the equations of his model.

When applied to this idealized universe, Einstein's model of gravity makes a simple but extremely disturbing prediction. *The model predicts that throughout the universe, the scale of length continuously changes as time passes.* (Note: There are actually two solutions to Einstein's equation, one in which the scale of length increases with time, the other in which it decreases. Fundamental physics does not distinguish between the directions of time, so anything that can happen in one time direction can also happen in reverse. For clarity, I'll refer only to the solution that corresponds to what we observe: the scale of length in the universe is shrinking as time passes!)

This strange behavior is illustrated in Figure 4.1.

The cube of space in the illustration represents a large volume of space in the model universe, filled uniformly with matter. Sup-

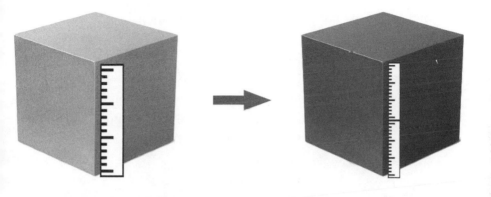

Figure 4.1. A large cube of space in a universe filled with matter. As time passes, the scale of length shrinks, as shown by the shrinking ruler. The cube of space now has a larger volume—but the volume is added inward, not outward!

pose that the cube starts out with a height of one "cosmic ruler." Einstein's model predicts that as time passes, the ruler will shrink. (That's what we mean by "the scale of length gets smaller.") Over time, the original cube of space will eventually have a height of two cosmic rulers, as shown in the figure. We interpret this to mean that the chunk of space will have twice the height as previously and will have eight times the original volume. Still later, as the scale of length shrinks even more, the cube of space will measure even more cosmic rulers in height.

Incredibly, Einstein's simple model predicts that the universe is continuously creating more space. *This new space comes into existence by virtue of the fact that the scale of length is shrinking.* There is no explosion hurtling the galaxies apart, no cosmic force. Just the silent, relentless creation of new space.

What would we observe if we were in this model universe? Let's say that our Milky Way galaxy, with us in it, is at one corner of the cube in Figure 4.1. Now observe a distant galaxy at the opposite corner. We would observe the distant galaxy moving away from us, for the simple reason that the distance between us is continuously increasing. The farther the galaxy, the faster it will be receding from our own, because the more space will come into being between us. That prediction from Einstein's model matches the observations that Edwin Hubble reported in 1929. The galaxies are like buoys bobbing in a great sea of space, serving as markers for the constant creation of new space.

You may wonder why I haven't drawn the two test galaxies at one corner of the cube. I did! But the cube in Figure 4.1 represents a chunk of space so large—at least one hundred million light-years wide—that the entire Milky Way galaxy would be smaller than a dot. You wouldn't see it in this drawing.

I haven't said anything about what happens to the scale of the test galaxies themselves. In this simple model, nothing; we are only justified in drawing a conclusion about the scale of space itself.[11]

It may come as a surprise that the expansion of the universe today is a very small effect. A chunk of deep space the size of the planet Earth expands by the width of one human hair over the course of a year. One human hair's worth a year! Who would ever notice such a tiny change? But *every* chunk of deep space expands by the same amount, and that tiny effect adds up when you consider the vast distances between the galaxies. As a consequence, galaxies are hurtling away from us at an impressive clip. For example, the Whirlpool galaxy is receding from Earth at more than a million miles an hour. And galaxies that are far enough away from us are receding faster than light! None of the light they emit will ever reach us, so they are beyond direct observation. We are stuck in a bubble of space that we call our "observable universe." It is huge, but it is finite.[12]

I want to call your attention to the other assumption of Einstein's model of the universe, that it is entirely filled with mass (that is, matter or other forms of energy, such as light). If that were not the case—if you merely had an enormous collection of galaxies surrounded by nothing but empty space—then gravity would indeed eventually draw all the galaxies together. So the recession of the galaxies is a phenomenon that applies to the universe as a whole, the universe-as-entity. The universe is more than the sum of its parts.

According to Einstein's simple model, the process of expansion gradually slows as matter becomes more diluted.[13] This dilution is indicated in Figure 4.1 by the darkening of the cube.

The Universe Expands Inward!

The term *expanding universe* has been in use for nearly a century, but it is misleading. People often ask, "What is the universe expanding into? How does new space push the existing space out of the way?"

The surprising answer: It isn't expanding into anything, because it isn't expanding outward at all! You may have seen illustrations comparing the universe to an expanding raisin loaf or to an expanding balloon. These illustrations are in error, because the before and after views are both drawn at the same scale. But that misses the essential point of Einstein's discovery: the scale of distance in the universe is changing with time.[14]

As you can see in Figure 4.1, the cube of space is not expanding outward. For want of a better term, it is fractalizing, or expanding inward. There is no limit to the volume of space you can create inside any boundary if you shrink the scale of distance sufficiently. Incredibly, space can expand infinitely, without ever stretching its finite boundary![15]

Note: For the remainder of this book, I will talk about the expanding universe, since that is the terminology everyone uses. But keep in mind that the universe is not expanding outward!

If you happen to have seen the old science fiction film called *The Incredible Shrinking Man,* you'll have seen a visualization of this idea. The hero of the film continually shrinks, thanks to the effects of radioactive fallout from 1950s weapons testing. At first he is a prisoner in his own basement. But as he grows smaller without limit, the basement becomes a universe to him, an entire cosmos within four walls. It is an imperfect analogy, but you get the idea.

How Difficult to Change Our View of the World!

If you find the scenario of an expanding or fractalizing universe disturbing, you are not alone. Einstein himself could not believe his own findings for many years. Had he believed his initial results, he would have made perhaps the greatest prediction in the history of science: he would have proclaimed that the scale of the universe changes continuously with time. It is never the same two moments in a row.

But that result seemed ridiculous to him. The heavens around us appear to be perfectly stable. The constellations of stars that were described and mapped by ancient civilizations thousands of years ago are the same ones that you see today. Even today we refer to the heavens as the *firmament,* from the Latin for "strong, steadfast, enduring." The stability of the universe was simply taken for granted.

So Einstein added a constant term to his equation that he thought would cancel out the instability of space and eliminate the prediction that the scale of length would change with time. The term he added is called the *cosmological constant*. (Some have simply called it a fudge factor, though technically it is not.)[16]

Einstein's 1915 paper was almost apologetic in tone. He offered to help the reader over the "rough and winding road" he had taken, "because otherwise I cannot hope that he will take much interest in the result at the end of the journey." He confessed that he "admittedly had to add an extension . . . [that is, the cosmological constant] which is not justified by our actual knowledge of gravitation. . . . That term is necessary only for the purpose of making possible a quasi-static distribution of matter, as required by the fact of the small velocities of the stars."[17] In other words, we don't see the stars moving rapidly, so we had better not have an expanding or collapsing universe. (Recall that, in 1915, it wasn't yet known that there are galaxies beyond our own Milky Way.)

Years later, Einstein would call his decision to modify his original equation "the greatest blunder of my life." But at the time, the implications of his theory were too bizarre for even him to contemplate.

Others were equally skeptical. The chairman of the Nobel Committee, Svante Arrhenius, gave a speech on Einstein's behalf in December 1922 in which he relegated Einstein's relativity to the status of philosophy. (Einstein had won the Nobel Prize, but not for relativity.) "There is probably no physicist living today whose name has become so widely known as that of Albert

Einstein," began Arrhenius. He continued by saying that while most of the discussion about Einstein's work centered on relativity, "this pertains essentially to epistemology and has therefore been the subject of lively debate in philosophical circles. It will be no secret that the famous philosopher [Henri] Bergson in Paris has challenged this theory, while other philosophers have acclaimed it wholeheartedly. The theory in question also has astrophysical implications which are being rigorously examined at the present time."[18]

One has the sense that the Nobel Committee was as cautious and perhaps dismissive of relativity as the secretary of the National Academy had been when he said he wanted to banish relativity to the fourth dimension.

But two other physicists were keenly aware of the monumental importance of Einstein's findings, and they would not be as hesitant as Einstein in proclaiming the expansion of the universe. One was Alexander Friedmann, a Russian physicist working at the Central Physics Laboratory in Petrograd, now Saint Petersburg. In 1922, Friedmann published a version of Einstein's model that did not include the fudge factor and that predicted an expanding universe. But the editors of the journal asked Einstein to provide a brief comment on Friedmann's paper. Einstein thought he detected a mathematical error and thus called Friedmann's result "suspect."

In reality, it was Einstein who had made the mathematical error. Friedmann got wind of Einstein's negative comment before it appeared in print, and he wrote desperately to Einstein for a correction. "The result I obtained does *not contradict* the case of a nonstationary world [that is, an expanding universe]," Friedmann wrote. "I earnestly ask you not to deny me an answer to this letter of mine, although I know how very busy you are going to be. . . .

In the event that you find the calculations in this letter to be correct, please do not reject my letting the editorial office of the *Zeitschrift für Physik* know about it; in this case, you might perhaps publish a letter of correction or make possible a printing of an excerpt from the present letter."[19]

But it was too late. Einstein's mistaken comment was published just a few days later. Although Friedmann would eventually become known as "the man who made the universe expand," he died just a few years later of typhoid fever in relative obscurity.

The Belgian mathematician, physicist, and Catholic priest Georges Lemaître fared better. In 1925, Lemaître used Einstein's model of gravity to investigate what would happen to a universe that was uniformly filled with matter. Unaware of Friedmann's earlier work, Lemaître independently came to the same conclusion that the scale of distance in the universe was unstable and would change as time passed.

Lemaître had the benefit of some of the early astronomical observations showing that the distant galaxies were getting farther away. So even though the theory's prediction was bizarre, he knew there was observational evidence to support it. In fact, Lemaître was able to use Einstein's theory to predict the relationship between a galaxy's distance and its speed, two years before Hubble announced his observations. As Lemaître put it in his 1925 paper (which remained untranslated until 1931), "The receding velocities of extragalactic nebulae [that is, galaxies beyond our own] are a cosmical effect of the expansion of the universe."[20]

Lemaître tried to persuade Einstein of his results at a conference in 1927, but Einstein would have none of it. "Your calculations are correct," Einstein is said to have replied, "but your physical insight is abominable."[21]

By 1929, Hubble and Milton Humason had published their astronomical observations showing that the galaxies are receding from Earth—the universe is expanding. But it took several more years for Einstein to drop his opposition to the idea of an expanding universe.

In April 1931, he returned to Berlin after a three-month visit to the United States in which he met with Hubble and other astronomers at the Mount Wilson Observatory. He had told the Associated Press that Hubble's observations "have smashed my old constructions like a hammer blow."[22]

In a paper entitled "On the Cosmological Problem of the General Theory of Relativity," Einstein eliminated his so-called fudge factor, the cosmological term that he had added to his equations in 1915 in order to cancel out the universe's expansion. In essence, he was allowing the universe to expand. He had no choice.

But he wasn't happy about it. Some insight into Einstein's discomfort can be gleaned from the paper itself. Considering the momentous impact of his theory, his paper has been called surprisingly "casual."[23] He dashed it off in just four days. He didn't bother to cite Hubble's pathbreaking paper containing evidence for the expansion, and for that matter he cited only Friedmann, whose work he had questioned just a few years earlier. "The first to try this approach," Einstein acknowledged, "was A. Friedmann, on whose calculations I base the following remarks."[24] Yet Einstein apparently didn't notice that some of Friedmann's calculations were in error.

The paper was to be among the last on cosmology that Einstein would write before going on to other scientific interests.

Today we can look back and marvel at the power of physics to predict such an extraordinary phenomenon as the expansion of the universe. And we think of scientific revolutions as suddenly changing our way of seeing the world. But scientists are by nature conservative and skeptical, as well they should be; extraordinary claims require extraordinary evidence. The actual process of change may be surprisingly slow.

At the time, Einstein was not certain about either his own predictions or Hubble's observations: Other interpretations of Hubble's data were possible. And Einstein's intuition had told him that the universe was static and unchanging; the predictions of his own theory were a challenge to his intuition.

And, to add to his chagrin, there was the problem of what the universe was expanding *from*. The equations of physics can predict backward in time, as well as forward. Einstein's model of the universe predicts that if you were to rewind the tape of the expanding universe, then all the matter and energy we see today would at some time in the past have coalesced at a single point.

The space we see now would have been squeezed to a point. The density of matter would have been infinitely high. The time line itself would abruptly end.[25] At that point, the equations of Einstein's model break down; they would no longer describe physical reality.

A situation in which physical quantities become infinitely large is called a *singularity*—something to be avoided at all costs. Einstein saw the singularity as a warning sign that his equations were no longer valid at that point. Someone else would have to improve and extend his model. It no doubt added to his discomfort at the predictions of his model.

Nevertheless, Einstein's model predicted that the universe must have expanded from a state unlike anything we see around us today. What was this early universe like?

5

The Big Bang and Beyond

The moment when our observable universe expanded from an extremely dense and hot state is called the Big Bang.[1]

The Big Bang took place right there where you are sitting. And it took place where every other living thing in the universe happens to be right now—because all those places were once part of the tiny volume of space in our infant universe. The particles of matter from which you are made, as well as those composing every other creature in existence, and every star and planet, were once shoulder to shoulder in the Big Bang. It was the reverse of America's motto, *E pluribus unum* (Out of many, one): out of one speck of matter and space came the entire observable universe.

The Big Bang is not a theory or a hypothesis or a guess. It is a conclusion supported by many lines of evidence painstakingly gathered over the last century. The phenomenon was predicted by Albert Einstein's model of gravity, which to date has survived every test thrown at it.

Much of our understanding of the Big Bang coalesced in the year 1948, in a rapid series of research papers by George Gamow, his student Ralph Alpher, and Gamow's colleague Robert Herman. Gamow was a Russian physicist who moved to the United States in 1934, after failed attempts to flee the Soviet Union by kayaking across the Black Sea with his wife. As a youth, Gamow had studied with Alexander Friedmann, and later at Göttingen, where half a century earlier the mathematician Bernhard Riemann had worked.

The three men—Gamow, Alpher, and Herman—were able to work out a consistent description of the first few minutes after the Big Bang. The infant universe was extremely dense and extremely hot, and it was expanding extremely fast.

Gamow originally thought that all the chemical elements had been created in the Big Bang, starting with simplest element, hydrogen. But Alpher calculated that the universe was expanding too fast and cooling too quickly for the higher chemical elements to be formed. He found that, after the universe cooled, the composition of the universe would be about three-quarters hydrogen and one-quarter helium (the next-heaviest chemical element). Only a trace amount of the next two elements, lithium and beryllium, would be created as well.

Alpher's calculations were borne out by observations. The predicted abundances are just what we observe in the gas that composes much of the galaxies and that forms the bulk of most stars. For example, our own star, the Sun, is about three-quarters hydrogen and one-quarter helium by weight. This agreement between theory and observation constitutes a second pillar of evidence for the Big Bang.

Alpher and Herman were also able to show that the Big Bang must have been extremely hot, and they predicted that the blazing light from the Big Bang should still fill the universe, coming to us from all directions in space as a faint glow. As the universe expanded, this light lost so much energy that it no longer has enough energy to stimulate our eyes. What was once visible light has become light in the microwave region of the spectrum, which can only be detected by special instruments.

In the late 1940s, however, many scientists were still skeptical about the reality of the Big Bang. In fact, the term *Big Bang* was coined derisively by the British astronomer Fred Hoyle as part of his radio show for the British Broadcasting Corporation. Even Edwin Hubble was said to have been skeptical about the Big Bang scenario throughout his lifetime.

The turning point came in 1964, when Arno Penzias and Robert Wilson, radio engineers at the Bell Telephone Laboratories in New Jersey, accidentally discovered the faint light from the Big Bang that Alpher and Herman had predicted. They were working on a telecommunications project to bounce radio waves off balloon-borne satellites. In an effort to eliminate all background sources of noise from their radio signal, they found a persistent hiss that appeared to come from all directions in the sky, continuously, day and night. (You can actually hear this static from the Big Bang for yourself when you tune your radio to a spot between regular stations.)

Modest and unassuming, Wilson said that he was too wrapped up in the technicalities of his work to fully appreciate its significance. That changed one morning when his father brought home the *New York Times,* laid it on the kitchen table, and opened it to an article describing the faint signal that his son had discovered. The reporter called the light "the afterglow of creation." Wilson says that's when he realized the import of his work. In 1971, both Penzias and Wilson were awarded the Nobel Prize for their discovery of this ancient light.

These three lines of evidence—the expansion of the universe, the abundance of hydrogen and helium, and the light coming to us from the infant universe—are three solid pillars of evidence for the Big Bang.

Einstein died in his sleep in 1955, so he never got to see the discovery of this third great pillar of evidence—the "afterglow of creation." But just as Helen of Troy's beauty launched a thousand ships, so did Einstein's physics launch a thousand experiments to study the universe's infancy. It is ironic that Einstein, who once reflected on his "marked lack of desire for direct association with men and women," should have spawned a century of scientific discovery, first with small teams of researchers, and eventually in large, multinational efforts to probe the earliest moments of our universe.

"I am a horse for single harness," he wrote, "not cut out for tandem or team work. I have never belonged wholeheartedly to country or state, to my circle of friends, or even to my own family. These ties have always been accompanied by a vague aloofness, and the wish to withdraw into myself increases with the years."[2]

The "afterglow of creation" has been studied with exquisite precision from locations as pristine and remote as the South Pole and outer space—most recently with space probes from NASA and the European Space Agency. A century of exploration has shown conclusively that everything we see in our observable universe was once packed into a region no bigger than a grapefruit.[3]

Today, we know more about what the universe was like fourteen billion years ago—and with greater confidence—than we know about what lies just a few miles beneath our feet inside the Earth.

Well, not quite. There was a surprise in store.

A Runaway Universe

As so often happens in science, a seeming paradox can point the way to progress. In this case, it was Einstein's model of gravity that both presented a problem and pointed the way to a remarkable discovery.

We've seen that Einstein's model of gravity predicts that the universe is expanding. But the model is quantitative as well, like an accountant's ledger. On one side of Einstein's ledger is the amount of mass in a big volume of space. The other side of the ledger tells us how fast that volume of space is expanding. The two sides of the ledger have to balance: the more mass there is, the faster space should be expanding.

But there's a problem, Houston. There isn't enough matter out there to account for how fast the universe is expanding. Astronomers can tally all the glowing galaxies of stars, all the dust and gas, and they can throw in an estimate for the planets, comets, and asteroids that are out there but are too dim to be seen at a distance, and, alas, they come up short.

In fact, there appears to be less than 5 percent of the matter needed to account for the expansion of the universe! Perhaps astronomers simply missed the vast numbers of frozen chocolate bars floating in interstellar space. How would we ever know they were there? Fortunately, there is an independent check: The amount of matter that was present at the Big Bang has a critical effect on the menu of chemical elements that emerged from that fiery event. If there were significantly more matter, the universe would have a different chemical composition from what it is observed to have now. So we know that ordinary matter really does account for only 5 percent of the mass that is out there.

But then what could possibly account for the 95 percent of the "missing mass" that we don't see? The answer is far stranger than what anyone would have expected, and it takes us into a new realm of mystery. It turns out that the universe contains two entirely new forms of mass. They are called dark matter and dark energy, and they are both important for our story.

Dark matter should probably be called invisible matter, because it does not interact with light, so it casts no shadows and is completely invisible. It is oblivious to the forces that hold atoms together, so it cannot form atoms; in fact, it passes right through ordinary matter. But dark matter, like every form of mass, exerts a gravitational tug on the world around it, and that's how we know of its existence. There is no "cloaking device" for gravity. If you exist, the world will eventually know about you through the gravity you exert.

Over the past half century, it has been shown that dark matter engulfs entire galaxies, including ours. The universe is filled with about six times as much dark matter as ordinary matter. Hundreds of thousands of particles of dark matter are passing through you right now. As of this writing, the nature of the dark matter particle is unknown.

Dark energy is even stranger. It is thought to be the mass of empty space itself. No one, of course, would think that the emptiness would "weigh" anything. How can space have mass if there is nothing in it? The idea originally bubbled up as an unwelcome prediction of quantum physics, which describes the world of the very small.[4] But that theory predicts such a large mass for empty space, nearly infinitely large, that physicists realized they must be missing something fundamental.

Remarkably, Einstein's model of gravity independently suggests that space might have mass. And it even describes the consequences of this mass, and how we might detect it and determine the amount as well.

A friend of mine once led a workshop entitled "How to Write a Poem That Knows More than You Do." Einstein's model of gravity is like that promised poem. It knows more than he did. Like a child, it has a life of its own; it is a wellspring of new meaning that continues to nourish generations of explorers. Einstein would never have thought that space could have mass, could actually "weigh" something. But his model accommodated the idea.

In fact, the model predicts that if space does have mass, then it would eventually cause space to expand much faster than ordinary matter would. There is a straightforward reason for this, which follows from Einstein's accounting ledger.

Previously, it had been thought that the universe's expansion should be slowing down. After all, as matter is diluted, its effect on space and time is gradually diminished. By contrast, the mass of space is not diluted as the universe expands, because new space brings new mass with it! As a consequence, the expansion doesn't slow down. (You may wonder how new energy can come into existence. Didn't we learn in school that energy can be neither created nor destroyed but merely transformed from one kind to another? Remarkably, the expanding universe has a loophole that preserves the law of conservation of energy.)[5]

In 1998, two teams of astronomers independently discovered that the universe's expansion really is speeding up—and by an amount that is consistent with the idea that space has mass.[6] The breakthrough is so far-reaching that in 2011 the researchers were awarded the Nobel Prize in Physics for their discovery. Their measurements, along with other lines of evidence, indicate that dark energy forms nearly three-quarters of the mass of the universe! And the nature of this mass, or energy, is completely unknown.

It is part of the ethos of science to try to disprove your own findings, the better to strengthen confidence in them. While other possible explanations for the accelerating universe are certainly being explored, the leading contender is that the phenomenon is caused by the mass of space itself. The "missing mass" now appears to be accounted for.

The mass of empty space is minuscule compared to anything in our world. A mere tablespoonful of sugar weighs about the same as a million trillion cubic miles of space. Yet the universe contains so much space that dark energy outweighs the effects of ordinary matter and even dark matter.

Why do we care at all about dark matter and dark energy? After all, they seem almost like abstractions, like ghosts in the night. It doesn't seem as though either of these components of the universe contributes to anything useful. You can't build an atom with them, let alone a skyscraper. So why should nature allocate 95 percent of the mass of the universe to these will-o'-the-wisps?

Remarkably, the universe apparently *needs* these exotica—and in just the right proportions—in order to create galaxies of stars and their planets and ultimately life. In the infant universe, dark matter provided the additional gravity needed to help matter coalesce into stars and galaxies. And the dark energy kept the pace of the expansion lively: Had the universe expanded too slowly, the heat of the Big Bang would have consumed the simplest element, hydrogen. Without hydrogen, water and life as we know it would not have formed. (You don't want to leave your batter in the oven for too long!) At the same time, if there had been too *much* dark energy, the universe would have expanded so quickly that galaxies of stars would never have had a chance to form. Matter would be too sparsely scattered through space to ever coalesce.

So, in a sense, there is a recipe for building a viable universe like ours. You can perhaps tweak the recipe a bit, but if you were to change the proportions of the starting ingredients significantly, you wouldn't wind up with an interesting place to live. This recipe is part of the universe's building plan, which we will explore in Chapter 6. And though we have no idea how the recipe was arrived at, it is not inappropriate to send your compliments to the chef.

What Happened before the Big Bang?

Although the Big Bang scenario is one of the great triumphs of science and reason, and although it describes an event that looks and feels like the origin of the universe, it still doesn't tell us what we really want to know: Where did all the matter and energy in the universe come from in the first place? What lies beyond the part of the universe that we can observe? How did time begin— if time even had a beginning? Why was the universe born with this particular building plan and not some other? No one yet knows.

Einstein's model of the universe is powerless to help. It arose from a retrospective view of the universe, a backward glance, anchored in what we observe now, informed by what we can deduce by running the "tape" of the universe backward in time and applying currently known physical laws. Unfortunately, that method is doomed to fail. Einstein's model of gravity breaks down as we approach time zero, because the universe would become infinitely dense as it shrank to a point, and Einstein's equations fail there. Furthermore, we know that the world of the very small is described by quantum theory, but Einstein's model of gravity has never been completely merged with quantum theory and made to accommodate its strange rules. In short, there is not yet a well-tested theory that can predict with assurance what the universe was like before the Big Bang, or what the underlying physical principles might be.

The Big Bang represents the boundary of our ignorance.

But there are theories in the making that take us before the Big Bang. One of the leading theories, called inflation, escorts us to a moment just before the Big Bang. In its current form, the theory postulates a mother universe from which our universe sprang. The mother universe is envisioned as an expanse of space possessing an enormous mass—perhaps similar to dark energy but much more concentrated. The mass of space in this mother universe is imagined to have been so large that—as Einstein's model of gravity tells us—even a tiny speck of space would have expanded to an inconceivably large size in an instant.

At the moment of the Big Bang, in some tiny region of this ever-growing mother universe, the enormous mass of space was converted into the fundamental particles from which our universe is built. In this scenario, the Big Bang was not an origin but merely a transition, in which preexisting energy was converted into matter and light. After the transition, the speck of space expanded at the less frantic pace we observe today.[7]

This mother universe is sometimes called the multiverse. There is no limit to the number of universes it can spawn. In fact, even when some portion of space transitioned to a "big bang," there would always be another portion that was expanding exponentially fast. In this multiverse, big bangs would take place at random locations and times, like flashbulbs lighting up an inconceivably large sports arena. The process would never stop.

Since space expands inward rather than outward, these multiple universes would not impinge on each other or compete for space. It turns out that each universe could be infinitely large and still fit unobtrusively next to its neighbors! The multiverse is an extraordinary prediction of continuous creation at a grand scale, with universes bubbling up from a primordial background of space and energy (portrayed schematically in Figure 5.1). It is believed that each one of the baby universes must have had an origin in time but that the whole assemblage—the multiverse—need not have had an origin and instead may be eternal.

The inflationary universe theory faces two big hurdles: No one knows the nature of the primordial energy that is assumed to drive inflation, or how it relates to the larger framework of physics. And there is not yet a successful theory that unites Einstein's theory of gravity with the quantum theory, which governs the world of the very small. Thus the inflationary universe theory is an ad hoc theory, and its predictions must be considered tentative and preliminary.

The possibility of multiple universes remains speculation. Many scientists disdain the idea, since there is no way to confirm the existence of these cosmic cousins. Nevertheless, the beauty of science is that, by testing a theory's predictions through experiments in our own universe, we can gain confidence in the plausibility of its predictions about regions we cannot observe directly.

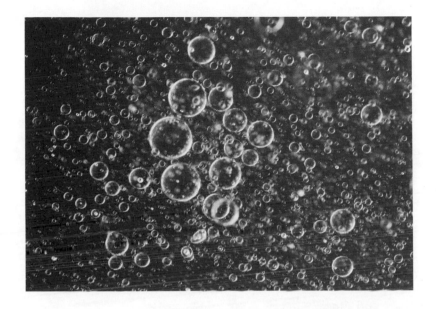

Figure 5.1. Schematic view of one conceivable multiverse. Bubbles of space expand at an enormous rate. Portions of each bubble may undergo "big bangs," creating individual universes in which the laws of nature may differ from our own. Credit: CC0 1.0 (detail).

A Universe Unbounded

Put aside for a moment all thoughts of physics. Put aside the strange ideas, the scale of distance, the dollhouse universe, the long debates, the sifting of evidence. Set aside our march through Einstein's "land of ideas," as Karl Schwarzschild described it from the battlefield. Put aside, if you can, the confusion that springs from compressing centuries of thought into these few pages.

Concentrate instead on the universe itself. Think for a moment just of the strange alliance of space, time, and matter—this bubble of creation that can reproduce itself, bring more existence into existence, expanding inward, filling in the infinitesimal as it grows. The universe is built in such a way as to create more of itself.

The universe has been a factory for space, time, and energy ever since its inception. This represents the first great stage of creation.

Think for a moment not of the immensity of the universe but of the single, undistinguished, undifferentiated beginning from which all of us derived. The Big Bang took place right here, where we live and work. During the Big Bang, the matter from which we are made was once side by side with that of our great-great-grandparents, cheek by jowl with that of our friends and our enemies, with the original particles of every living creature in the universe—all those who are now separated from us forever by space and time.

But as marvelous and mysterious as our cosmic origins may be, it is the unfolding of the universe that gives it meaning. For we live in a bubble of creation, and not for a single second has the universe been idle.

Think for a moment not of the matter from which we are made but of the strange substances called dark matter and dark energy, which dominate the universe, and whose proportions—so fortuitously specified by nature—seem to suggest a universe that knows what it is doing.

Does it? It is time to take a closer look at the universe's building plan.

6

Building Plans

When the universe was born, it came with an infrastructure—a set of building materials, rules, and specifications that guide the development of the universe.

The infrastructure includes the basic building blocks of matter, such as the particles from which atoms are made.

It also includes the basic forces that animate matter, such as gravity, electromagnetism, and the nuclear forces. These forces are the heart and soul of the universe (Latin: *ex animo,* "from the soul"). Without the forces of nature, the building materials would just sit there and do nothing.

No one yet knows why the forces of nature have the relative strengths that they do. For example, electromagnetism is about a trillion trillion trillion times stronger than gravity. That's why it takes the entire Earth to make a paper clip fall to the ground, yet a tiny child or magnet can easily pick it up again. Gravity rules large conglomerations of matter—planets, stars, galaxies, the universe as a whole—while electromagnetism dominates our everyday world of molecules.

The infrastructure also includes a set of rules that we often call the "laws of nature" but that are merely our current, best description of how nature behaves. As we learn more, the laws get revised.

Since we don't have access to an original architect, we can't ask directly, Why these specifications and not some other set? But one thing seems to connect them all: the infrastructure of the universe makes it possible for life to emerge. I'll be more restrictive: without these particular specifications, even the simplest atom would not have been created.

The hydrogen atom is probably not on your list of treasured possessions. It is the simplest of chemical elements, and it has just two moving parts: a single proton that sits in the center, and a much lighter electron that orbits around the proton. As matter goes, hydrogen is as unassuming as you can get. Yet the universe has endowed this building block with unique properties that underpin the entire development of the universe.

Remarkably, every hydrogen atom in the universe is identical to every other. Not just "sort of the same," or very similar—absolutely identical. The atoms can occupy very different environments, of course. And they can be broken apart into their constituents. But when they are reconstituted, they are again indistinguishable from all others.

Furthermore, atoms do not age. Each one of the thousand trillion trillion hydrogen atoms in your body is nearly fourteen billion years old, a relic of the Big Bang itself. Compare that with the building blocks in a child's LEGO set; they may all seem identical, but come back in a few billion years and see what they look like then!

The extraordinary uniformity of hydrogen atoms is true of other chemical elements as well, such as oxygen or iron. And it remains true when atoms combine to form molecules. Thus a molecule of vanillin in your vanilla pudding will be identical to a vanillin molecule in some alien's pudding in some galaxy on the far side of the universe. (It may not taste the same to an alien, but that is no fault of the molecule.)

This uniformity is essential to the development of the universe. The extreme complexity of life itself relies on the extreme simplicity of its atomic building blocks. *Order could not have arisen if the infrastructure of the universe were not orderly.*

Another feature of nature's building plan is that space has three dimensions. (You can move up and down, forward and backward, and left and right.) It didn't have to be this way. You can picture a one-dimensional universe, like a string; or a two-dimensional universe such as a flat sheet of paper; and there can be any number of higher dimensions of space. But it turns out that only in a three-dimensional world do stable orbits exist. So neither atoms nor solar systems would exist without this specification.[1]

Nature has also ensured that atoms have a simple, clearly defined internal structure. The possible structures that an atom can have are nothing at all like a solar system. In an atom, the electrons can only occupy a few specific orbits. Contrast that with planets orbiting a star: planets can have a vast variety of orbits. In fact, no two planets orbit their stars in exactly the same way; every orbit is slightly different. Some planets may be closer to their stars; some may be more distant. Some may have circular orbits; some may have greatly eccentric orbits that carry their planets first closer to and then farther from the central star. An orbit may change with time as a passing object perturbs the planet and diverts it from its course. A planet can spiral into its star and be lost forever. In short, planets carry evidence of their histories, like everything in our large-scale world.

Fortunately, atoms are governed by nature's quantum rules of behavior. The rules ensure that electrons can only occupy a few, very specific orbits within the atom. The rules prevent electrons from spiraling into the proton and collapsing the atom. *Thus nature's specifications guarantee the stability of atoms.*

Furthermore, nature has arranged that no more than two electrons can occupy a single orbit in an atom. Nothing like this rule exists in our everyday world. For example, you could launch a hundred space stations into the same orbit, at the same distance from Earth. But within an atom, only two electrons can occupy the same orbit. Without this rule, this orderliness, this predictability, atoms would not form molecules with well-defined, dependable properties.

Nature strictly enforces this rule by imposing a bizarre property on the electron: When an electron is rotated a full turn around (360 degrees), *it is no longer the same object as when it started;* it actually has different properties. You must turn it an additional 360 degrees for the electron to have the same properties as when you started![2] There is no object in our everyday world that behaves this way.[3]

Many students find quantum theory to be difficult going, almost an imposition, because the rules are counterintuitive and are utterly different from the way our everyday world works. The rules of the microscopic world are not how we think the world ought to behave. The quantum world is often thought of as a "fuzzy" realm, thanks to phrases such as "the uncertainty principle." But when it comes to atoms, the quantum rules enforce a uniformity and precision that has no equal in our lumbering large-scale world. The rules are not merely a description of nature; they are a guarantor of the stability of the world.

Here is another of nature's building specs: the building blocks of matter all have very specific masses. For example, each proton in the universe is 1,837 times more massive than each electron. A neutron weighs more than a proton plus electron. But what would happen if nature had specified some other values?

Why These Building Plans?

Researchers have investigated what the universe would be like if you could tweak some of nature's building specs. If you made gravity a little stronger, or electromagnetism a little weaker, what would happen? If you put the neutron on a diet so that it weighed less, what would happen?

It turns out that there is not much wiggle room—if any— in nature's specifications. If you and I were to go on a diet and lose a few pounds this week, it might make us happy, but it certainly wouldn't alter the arc of our lives. But if a neutron lost the same proportion of its weight—say, about 2 percent—the world would be very different, and you and I wouldn't be in it.[4]

In the same way, even a slight tweak of the force of gravity or electromagnetism has large consequences that typically preclude life. Atoms become unstable, or stars evolve too fast or too slowly, or the universe expands at the wrong rate. Even the proportion of dark energy is critical (see Chapter 5). If there is too much dark energy, the universe expands too quickly and matter never has a chance to clump and form stars, galaxies, planets, and life. If the universe expands too slowly, all the hydrogen atoms get cooked into other elements. Without hydrogen atoms, there would be no water—and you know what *that* means.

So why is the infrastructure of the universe so hospitable to life? How were these specifications made in the first place? This problem is known as the fine-tuning problem. It is one of nature's deepest mysteries.

Three responses to these questions have been widely discussed, and all of them are speculative. The oldest response, of course, is the theological perspective; it has so many facets reflecting so many cultures that I could never do them justice here. To mention just one example, Isaac Newton wrote that God kept the world ordered and might even intervene from time to time—for example, to correct the orbits of comets. Gottfried Leibniz scoffed that God would have created a perfect universe and would surely not need to wind the universe's "clockwork." In many theological perspectives, humans figure prominently as the intended end product of the universe's development.

A second response is the scientific perspective that has held sway for the past few centuries. This perspective assumes that there is a logically consistent, simple framework for understanding why the universe must be built the way it is, and that we can progressively uncover this basic truth through theory, observation, and reasoning.

This approach has been enormously successful in the past. Physicists have assembled a deep understanding of the universe on the basis of ever-fewer basic ideas. Unfortunately, when it comes to making progress on the fine-tuning problem, physics seems to have hit a brick wall. The well-tested frameworks of physics have not yet provided an explanation for why the physical universe is constructed the way it is.

In this scientific perspective, the universe has no stake in its products; nature does not "intend" to produce either atoms or life.

However, it is noted that the fine-tuning problem exists only because intelligent life has discovered it. If nature were not just so, we wouldn't be here to ask the question—an observation sometimes called "the anthropic principle."[5] So intelligent life is a factor in the problem, though not an explanation.

There is a third response, based on the idea that our observable universe may be just one region of a much larger multiverse (see Chapter 5). In these other regions of space and time, the building plan need not be the same as our own. There might be quite different building blocks of matter, forces with different strengths, different rules, and even a different number of dimensions.

Driving this scenario is string theory, which is a leading contender for the next generation of physics theories. String theory predicts that there may be a staggering number—nearly an infinite number—of possible ways to build a universe, so many that it would be impossible to list even the barest fraction of them.

Taken together, the multiverse and string theories describe a vast landscape of universes, almost all of which would be stillborn, and in which atoms and life could not exist. Most would flicker into and out of existence in a flash. But, very occasionally, there would be a flourishing universe like ours.

In this perspective, the fine-tuning problem seems to vanish. The building plan didn't have to be the way it is. We just got lucky. In fact, this perspective has spawned the catchphrase "accidental universe." It is a misleading phrase, because even in this scenario, our universe would not be accidental. It would just have many siblings. (I think of a mother going through the rigors of in vitro fertilization and winding up with quintuplets. To call any of her children "accidental" would seem inappropriate; the process happened to produce all of them.)

As striking as this scenario may be, it is not yet supported by evidence, and both string theory and the multiverse are still on the drawing board. For example, it is not known whether all the building plans that are possible in string theory could actually be implemented in a real, physical universe. There may well be some underlying principle that ensures that only *our* universe's plan could result in a physical universe.

A Fourth Perspective

John Archibald Wheeler's 1983 question provides a fourth perspective on why the universe's building plan is so hospitable to life.

Wheeler's question was motivated by the puzzling connection between the laws of physics and the presence of an observer. It was a problem that had long plagued quantum mechanics, the theory that describes the microscopic world.

It had been shown a few years earlier, for example, that quantum theory was not able to describe the evolution of a model universe unless you first divide the universe into two parts: an observer and the rest of the universe. It was a strange predicament. Here was a set of mathematical tools—quantum theory—that was able to accurately predict with exquisite precision the behavior of atoms and molecules. Yet the theory seemed dependent on the participation of an observer. And what constitutes an observer? Does it have to be intelligent and aware? Could it be a turtle? A machine? To this day, the problem remains a riddle that is not entirely solved.

Another enigma is the concept of now, the present moment in which you and I spend our lives. We may regret the past and look forward to the future, but every second of our lives seems to lie in the fleeting instant of time that we call "now."

Strangely, physics has no concept of now. (As Einstein put it when trying to console the widow of a dear friend, "To a practicing physicist, the past, present and future are but one and the same.") How odd that physics should have such profound insights into time as to be able to predict the future and the past and yet have no way of distinguishing what we call the present. In fact, it was predicted by Einstein and confirmed by experiment that one person's now can be another person's past or future; there is no universal now. So the very notion of the present depends on who is doing the observing! Through our awareness of time, we help to give meaning to time itself.[6]

For these and other reasons, Wheeler imagined that intelligent life might be part of the universe's agenda. It would be a kind of mutual pact, with each party helping to bring the other into existence.

In this perspective, life and awareness are as fundamental as nature's building plan. We are biased against this idea, because we assume that the plan was influenced only by events that preceded it. Yet the plan seems prescient, for as we will see, it relentlessly shapes both the conditions for life and life itself. It is as though the universe were constructed of a piece, without regard to time.

Paradoxical? Perhaps, but no stranger than the behavior of a simple electron. An electron moves over time like a wave: it takes many paths simultaneously. Yet when an electron is observed at a single instant, it is always a particle; it is never spread out in space. The universe, too, may demand a dual perspective: one that unfolds in time and one that does not.

The physicist J. Robert Oppenheimer noted that "two ways of thinking, the way of time and history and the way of eternity and of timelessness, are both part of man's effort to comprehend the world in which he lives. Neither is comprehended in the other nor reducible to it. They are, as we have learned to say in physics, complementary views, each supplementing the other, neither telling the whole story."[7]

Even though Wheeler was a consummate physicist, he realized that in the last analysis, the universe is not simply a set of equations. Many years ago, at the end of a talk on cosmology that he gave at Harvard University, he scribbled on the blackboard an equation that no one had ever seen before. "Imagine," he said, "that this equation explains every phenomenon in the universe, that it is the ultimate description of our world." Then he waved an imaginary magic wand and shouted, "Fly!"[8]

The equation just sat on the blackboard. "Fly!" he shouted again and again. Still, the equation just sat there.

"The truly amazing thing about the universe," he concluded, "is that it flies." Physics provides deep insights into the world. But its description of the world is not the same as the existence of the world. The universe is not an equation, not a description, not an idea. The universe *exists.* It gives form and substance to mere ideas. And while the search for our cosmic origins remains one of the most exciting of all adventures, the universe's mystery lies not just in its origins but equally in its forward motion.

The universe is in flight. Where is it going, and what has it been doing? The universe indeed has a story with a marvelous trajectory. And that's where we are headed in Part 2.

Part Two

HOW DID STRUCTURE ARISE
FROM CHAOS?

7

An Apple Pie from Scratch

On the last day of school in third grade, a stranger came to our classroom and sold every child a packet of seeds for two cents. The packet was just a small brown envelope with two words: Morning Glory. I planted the seeds in an old pot on the window ledge of our eighth-floor apartment and promptly forgot about them.

When I returned to the city at summer's end, I had still forgotten about the seeds, until the landlord called. Were we responsible for the flowers that were cascading down to the fourth floor, some forty feet below, blocking the neighbor's view?

To this day, I remember my astonishment that a tiny, unpromising seed could produce such a cascade of color—and all with the help of just a little rain, soil, sunshine, and fresh air. If you have ever watched a plant grow from a seed, you will have shared the same wonder. How does a seed know how to create such form and beauty? What kind of instructions does it have, and how are they packed inside that tiny seed?

As strange as it sounds, the universe itself can be considered a kind of seed. In its infancy, our observable universe was small, round, and densely packed with matter, just like the morning glory seed. And it came with its own set of instructions—the building plan—that guided its future development.

In fact, the universe is the ultimate seed, the überseed, because it is completely self-sufficient, developing without the help of any outside materials at all. The universe creates its own sunshine and water and soil and, given enough time, its own little boys and girls to plant flower seeds on windowsills. You have to be patient, but the results are spectacular.

Defying Entropy?

There is something deeply troubling about this. We can at least fathom how a morning glory seed might create a living sculpture of flowers and leaves. After all, the seed is filled with complex, living cells, each of which contains detailed instructions in the form of DNA. These instructions have been honed over a billion years of earthly development. The seed already knows how to harvest the raw materials around it, and how to gradually incorporate them into itself.

But the infant universe started with no experience and no structure on which to build. It was not a finely honed machine, like a living cell. It was jumble of disorder and chaos. In fact, the universe seems to defy one of the fundamental laws of nature: the increasing entropy of the universe.

We learn in school that in any natural process, entropy always increases: "Everything tends to become less organized and less orderly over time."[1] That is a popular statement of the second law of thermodynamics, formulated in the nineteenth century.

Entropy has widely been taken to be the same as disorder, and that's the way we use the word in everyday speech. At first glance, this seems perfectly reasonable. A vase falls to the kitchen floor and shatters into a hundred pieces, a disorganized mess; those pieces will never spontaneously organize themselves back into the vase again. The Pyramids crumble and decay over time; they will never spontaneously reassemble themselves. Entropy. The world seems to tend inexorably toward disorder.

Yet the story of the universe is the opposite of this. The universe has progressed inexorably toward more order. (See Figure 7.1.) Observations over the past few decades have confirmed that the universe *started* as the most disordered state of matter, a ball of hot gas. This gas had no discernable structure; it was uniformly spread throughout space to within one part in one hundred thousand—far smoother than the smoothest chocolate mousse. The temperature was the same everywhere to within one part in one hundred thousand—vastly more uniform than the temperature of the air surrounding you right now.

And yet from this simple, featureless, unpromising state arose all the complexity and beauty of the universe. Every star and planet and galaxy arose from this disorder. Every cloud, every raindrop, every melody, every thought and idea arose from this disorder. Flowers cascading from windowsills arose from this disorder.

So, if everything tends toward greater disorder, then how in the world did we get here? This paradox is illustrated in Figure 7.1, which shows a patch of the universe as it looked almost fourteen billion years ago, compared to the same patch of space today. We'll explore this question from several perspectives. But first we need to resolve the conundrum of entropy.

Figure 7.1. *Top:* A patch of the universe as it looked 13.7 billion years ago. *Bottom:* The same patch of the universe today. If "all things tend toward greater disorder," where did the marvels of the universe come from?
Credit: Photograph by the author.

Then we will be able to appreciate the unfolding story of how nature got us here.

"Print the Legend!"

There is a simple resolution to the paradox of a universe that went from disorder to order: it may surprise you, but entropy is not the same as disorder.

In the 1962 John Ford film *The Man Who Shot Liberty Valance*, the meek James Stewart becomes a legend for having shot the villain. But—spoiler alert—he really didn't; John Wayne did. The tale of Stewart's accomplishment has been repeated so many times that it is accepted as fact. Even when the truth comes to light, the newspaper editor continues to print the legend. "When the legend becomes fact," the editor explains, "print the legend!"

The idea that entropy is disorder long ago became its own legend. One of the early pioneers of entropy, the German physicist Hermann von Helmholtz, defined entropy that way in 1883. "One may characterize the magnitude of the entropy," he wrote, "as the measure of the disorder."[2] That has been the popular go-to definition of entropy ever since. (For example, the physicist George Gamow wrote that "the Law of Entropy can also be called the Law of Increasing Disorder."[3] The same definition pops up in dictionaries of physics: "Entropy can be thought of as the disorder of a system."[4] And that's what it says in the National Science Education Standards, which mandates what high school students should learn.[5])

However, in the late nineteenth century, the German physicist Ludwig Boltzmann showed that *the entropy of a system is actually a measure of two different properties.* One part of entropy is indeed what we commonly call disorder. The other part is a measure of the heat generated by the system. Both parts contribute toward the entropy.

It is possible for a system to spontaneously become *more* ordered, provided that this decrease in entropy is more than compensated by the heat generated in the process. In nature, this typically happens when a force of nature is at work—such as gravity or electromagnetism—because when a force acts, it typically generates heat. A few examples will illustrate this.

First consider an example in which no forces are involved. When you open a bottle of perfume, the fragrance molecules escape from the bottle and waft through the room. "What a lovely fragrance," people on the other side of the room will say. The system certainly becomes more disordered as the fragrance molecules migrate. Originally confined to a small space, the molecules end up moving helter-skelter everywhere. The entropy increases during this process, as it must. Furthermore, those fragrance molecules will never find their way back into the bottle, no matter how long you wait. The increase in entropy is a one-way process.

But suppose that instead of fragrance in a bottle, we look at a giant cloud of gas in deep space. This system of molecules is so large that now the force of gravity becomes a factor. The molecules cannot wander off into the lonely reaches of outer space. Instead, they fall toward each other under the influence of gravity. This is how our own Sun formed, and how stars continue to form today. So here we have a case in which the system becomes *more* structured of its own accord. It is as though those fragrance molecules were suddenly finding their way back into the bottle. The difference here is that a force is at work—gravity.

The process generates a great deal of heat, because the molecules gain energy as they fall. The heat produces more than enough entropy to offset the ordering of the system. So the entropy of the system increases even though the system becomes more ordered. Fortunately, nature provides processes for radiating some of this heat into space, which helps the stars coalesce.

There are similar examples around your home. If you put a drop of ink into a glass of water, for example, you know that the ink molecules will quickly diffuse through the water. The system becomes more disordered and more uniform. No matter how long you wait, the ink molecules will never reassemble into a drop of ink again of their own accord.

But suppose you shake up a mixture of oil and vinegar for a salad dressing. You mix them as thoroughly as you can. Now there is a force at work: electromagnetism. The droplets of water are attracted to each other much more strongly than they are to the oil. So, eventually, the two substances will separate back into oil and water.

Again the system spontaneously becomes more ordered—by separating into a layer of oil and a layer of water—as a force of nature goes to work.

Defeating the Heat Death?

When we examine the story of the unfolding universe in the next chapters, we will see how the forces of nature act to shape the universe. There is a catch, however. As the forces work their magic, they generate entropy in the form of heat. And, as the scientists of the nineteenth century quickly realized, that is a scenario that cannot go on forever. For what happens when the forces' ability to do useful work has been converted to heat?

In the middle of the nineteenth century, a chilling description of the end of the world was put forward by the scientists who had pioneered thermodynamics and pondered its implications. They predicted that the universe would eventually run down as all the useful energy was used up, leaving only heat in its wake.

Helmholtz described the "heat death" of the universe, a time when all energy would be transformed into heat at a uniform temperature, and all natural processes would come to an end. "The universe from that time forward," he said in a lecture in 1854, "would be condemned to a state of eternal rest."[6]

In the same year, Lord Kelvin (William Thomson) predicted at a meeting of the British Association that "the end of this world as a habitation for man, or of any living creature or plant at present existing in it, is mechanically inevitable."[7] Even the physicist who had coined the term *entropy,* Rudolph Clausius, pointed out that once the entropy approached a maximum, "the occasions of further changes diminish." And when the entropy of the universe finally reached a maximum, "no further change could evermore take place, and the universe would be in a state of unchanging death."[8]

If we could transport these giants of science into the twenty-first century, we could put their minds at ease. As we are about to see, "eternal rest" is not yet on the universe's to-do list. At least not anytime soon.

From Chaos to Apple Pie

"If you want to make an apple pie from scratch," the astronomer Carl Sagan once said, "you must first invent the universe."

It is easy to make an apple pie; all you have to do is follow the recipe. It is a lot more difficult if you want to make the pie from scratch. Where did the apples come from? Where did the wheat come from? Where did *we* come from? What challenges did the universe surmount in order to go from chaos to apple pie? You may think that creating a universe is the profound part of the story, but it turns out that getting to the pie is at least as miraculous.

Our universe took nearly fourteen billion years to arrive at an apple pie. It might have been able to shave a few years off here and there, but as we'll see, it required pretty much all of that time to get the pie ready.

In the following chapters, we'll explore four highlights from the long march to an apple pie. Each one involves a hurdle that the universe faced in creating the pie, not to mention the rest of the world. In exploring each episode, we will see the marvelous ways in which nature managed to surmount the hurdle— further evidence that somehow, in some way, life is on the universe's to-do list.

From the scientific perspective, several themes will come to the fore as the story unfolds. One is that nature thinks outside the box, almost literally. That is, nature ensures that matter and energy are not perpetually confined within a box. From the universe's point of view, a box is a coffin; anything inside will eventually come to equilibrium—and equilibrium is incompatible with life. Even the universe itself is not a box, since it continously creates new space.

A second strand of the story is that the four fundamental forces of nature each take center stage in succession. (The forces are gravity, the strong nuclear force, the weak nuclear force, and electromagnetism.) The four forces of nature are always in play simultaneously, of course, but nature has arranged for each to dominate the stage in a specific sequence. Each force has a critical role in shaping the apple pie.

Another strand is the clever way in which nature uses the so-called waste heat produced as each of the forces does its work. This heat turns out to be essential for the development of the universe.

So while it is true that "useful energy" is continually being converted into heat, that heat is itself useful.

Perhaps the strangest aspect of the universe's forward motion is that it has a story at all. Each step in the unfolding universe is built on what has come before yet seems to anticipate what is still to come. The universe is not a mere conglomeration of facts. How effortlessly it seems to blossom, just like that morning glory seed.

8

Into the Abyss

If you could magically drop in on the universe four hundred thousand years after the Big Bang, you would be appalled.

There was no place to sit down. There was nothing from which you could fashion a chair, even in principle. There was nothing to shelter you from the intense heat and blinding light, and no way you could create such a shelter, even in principle. A glass of water was out of the question, since the very concept of water didn't exist yet: there was no oxygen to make H_2O, or silicon to make glass.

The universe was filled with only the two simplest elements, hydrogen and helium. Neither was of use yet to anyone. Hydrogen had nothing to combine with, and helium is the least reactive of all elements.

There was no up or down, no orientation or discernible direction of any kind. And of course there was no need for a map, since there was no place to go. Every location was the same as every other.

It was a universe devoid of structure. Nor was it obvious that there ever would be any structure. At the time, no one could have discerned the course on which the universe had embarked.

If the universe had been enclosed in a box, even an infinitely large box, no structure would have formed. The story would have ended there. The universe would have found its "eternal rest" as a remarkably boring soup of hot gas. *It would be the end of the world, not the beginning.* So it is fortunate for us that gravity took center stage in the early universe.

Gravity set the universe on its proper course in two ways, each of which had to be in balance with the other. First, recall that gravity is responsible for the expansion of the universe (see Part 1). New space continuously streams into existence, and it is doing so even today. The expansion is critical to our story, because it enabled the universe to cool, just as any gas cools when you give it more room. An extremely hot universe would not have allowed the formation of structure.

As the infant universe cooled, it went dark. Within a million years after its birth, it was deeply, fundamentally dark. You could not turn on a light, even in principle, since there was nothing from which to fashion any kind of light, and there was no source of energy that could be harnessed to produce light.

Space was clear for the first time. Deeply, profoundly clear. But there were no stars to pierce the perpetual night, no landmarks, still no up or down. And no stars would ever have formed had the universe not solved two more challenges. The first of these challenges was the extraordinary uniformity of the universe. As mentioned, matter was uniformly distributed throughout the infant universe to within one part in one hundred thousand.

Had the universe been *completely* uniform, no stars would have formed. Just as a raindrop needs a speck of dust around which to coalesce, so does matter need some preexisting clump to which it is drawn by gravity. If the universe had been completely uniform, it would remain uniform as it expanded. It would be just a question of time before matter would be too thinly distributed through space to ever form stars.

In our own kitchens, we try to get batter as smooth as possible, with no lumps. But in nature's kitchen, when you are making a universe, you need a slightly lumpy batter if you are ever to form galaxies, stars, and planets. Thanks to the quantum behavior of nature, the universe had a slight lumpiness from the very beginning. This texture is built into the infrastructure of the cosmos.

You can see the graininess of the baby universe in the image in Figure 8.1. Previously, we saw that the baby universe looked uniformly white. In this image, the contrast is enhanced almost a thousandfold to emphasize that the universe was not completely uniform when it was born. The brighter and darker regions correspond, respectively, to regions that are very slightly more dense and very slightly less dense. The image was taken from the Planck space-based observatory, a joint project of NASA and the European Space Agency. Remember, this is literally a photograph of the past; the light has taken so long to get to us that it shows us what the universe *used to* look like. By now, the denser regions will have become galaxies of stars and their planets, similar to the Perseus cluster of galaxies seen in Figure 8.2.

It is marvelous that the same set of quantum rules that dictates the complete uniformity of atoms also dictated the nonuniform distribution of matter in the newly born universe. The universe relies on both the precision of atoms and the haphazard way that the first matter was distributed through space.

The baby universe faced another serious challenge. Even with slightly lumpy regions, it still takes time for matter to coalesce. The infant universe did not have all the time in the world, because

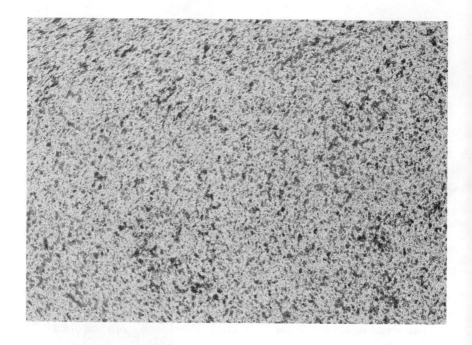

Figure 8.1. Light from our infant universe. Observations made by the Planck spacecraft reveal a very slight graininess (contrast enhanced). The lighter areas are slightly hotter, denser regions of space. Credit: NASA/ESA.

it was expanding rapidly. If the universe had expanded too quickly, matter would have been diluted before gravity had a chance to draw it together into coherent structures. No stars would ever have formed.

And if the universe had expanded too slowly, it would have remained hot for too long. In that case, the intense heat and pressure would have cooked all the hydrogen atoms into heavier elements. Without hydrogen, life as we know it could never have emerged. Hydrogen is required to make water.

Figure 8.2. The universe today. The denser regions in the infant universe eventually gave rise to clusters of galaxies, such as the massive cluster shown here. Credit: NASA / ESA / HST Frontier Fields Team (STScI).

Once again, the universe's infrastructure guaranteed that things would work out properly. It didn't have to be that way. One could imagine a universe with a significantly different amount of matter or dark energy, or a different strength of gravity; but in that case, we wouldn't be here. Again, the universe was built from the start with a clever set of plans.

Scientists have confidence in this story because they have well-developed mathematical models describing the process, they have superb observations of the early universe, and they can use super-computers to simulate the universe's evolution and then compare the results to observations. These simulations graphically reproduce the formation of structure in the universe, including the evolution of the first galaxies and clusters of galaxies.

In short, gravity was able to work its magic at two different scales: at the scale of the whole universe—the universe-as-entity—by causing it to expand; and at the local scale, by attracting matter to the neighborhoods with slightly more matter to begin with. Now the universe glittered with the first stars and the first galaxies of stars. The night skies were filled with glorious starlight. Gravity was beginning to endow the universe with shape.

But alas, even after a million years, there was still no vacation spot. There were not yet any materials to form planets like Earth. There was no place to sit down, no vantage point from which to gaze out on creation. It was still a hydrogen and helium universe. The cosmos had a lot of work to do.

The next challenge for the universe was daunting: "Make something of yourself!"

9

Into the Cauldron

On a clear summer morning in the year 1054, an artist from the Anasazi people in the American Southwest observed one of nature's rarest and most startling sights. A light appeared in the sky, so bright that it joined the Sun and Moon throughout the day.

The Anasazi vanished for reasons unknown and have thus been lost to history. But a record of their sighting remains both as rock paintings and as decorations on pottery. A drawing with twenty-three rays may represent the number of days that the strange light was visible. We don't know for sure what the Anasazi drawings depicted, but it was probably the same event recorded by Chinese observers on July 4, 1054—a "guest star," brighter than any other in the sky, lasting for twenty-three days.[1]

The ancients had witnessed one of nature's greatest events: the explosion of a star. Stars live and die just as we do—and it turns out that our own lives depend critically on theirs.

At first glance, stars might seem like uninteresting objects. A star is a nearly featureless ball of hot gas that seems to change little if at all for millions or even billions of years. But once you become familiar with a star's properties, it is hard to see it as anything short of miraculous.

Stars are the engines that create the chemical elements. They are the great alchemists of the universe, transmuting the two simplest elements, hydrogen and helium, into the heavier elements from which our world is made. Much of the work goes on deep in the interior of the star. That is where the nucleus of a hydrogen atom (a single proton) is fused with others of its kind. The process gradually builds up the elements, from helium to beryllium to carbon to nitrogen and so forth. The larger the star, the further this process goes and the more kinds of elements the star can produce. There are various, complex routes in this manufacturing process, but all of them are driven by the nuclear force, which is the strongest force in nature.

Star Diving

Let's take a journey deep into one of the stars on which our lives depend—a star that created the elements in the ground under our feet, in our own bodies, and in the apple pie. To do this, we'll have to go back in time a few billion years, when the star lived. And we'll have to dive deep into its core, where the temperature is millions of degrees and the pressure is a billion times that of the deepest ocean floor.

As the nuclear force acts inside a star, it generates tremendous heat. Nature does not waste this heat; in fact, it is essential to the formation of successively heavier elements. The reason is that in order to fuse two nuclei together, you have to bring them very close to each other. This is not easy to do. (Each nucleus is electrically charged, and as you may have experienced, like charges repel each other—the phenomenon that causes flyaway hair.) Very high temperature and pressure are needed to fuse the heavier, highly charged nuclei.

The star's core is a cauldron of creation—a kind of infernal Santa's workshop without the elves—where the first of nature's great gifts are created: the chemical elements from which everything in our world is fashioned.

Eventually, each chemical element will weave its own subplot in nature's story. In the cauldron, for example, are made sodium and chlorine, which will one day salt the oceans of Earth, along with our sweat and our tears. Also made here is silicon, which will fill every grain of sand on every beach, grace the world's fine china, and empower every computer on the planet.

And here is iron, the most stable element, which one day will stiffen the steel spine of skyscrapers, stain the Grand Canyon and the planet Mars red, and bind the oxygen in every red-blooded person's blood. And here too is copper, which will make a crustacean's blood run blue, and which will go on to carry the electric power that lights our cities. And there is the magnesium that will give the green to every chlorophyll molecule in every leaf, including the leaves of the apple tree, which even now is preparing fruit for an apple pie.

There are entire worlds waiting to be formed from these chemical elements. Planets, cities, apple trees. You. There is just one problem: the elements are locked inside the star. If the star were merely a factory, then the elements would remain forever trapped in the star's interior. There would never be an apple pie, nor any living thing to bake it.

Fortunately, a large star is designed to do more than merely create chemical elements. It is also designed to destroy itself.

When a large star runs out of its fuel, after just a few million years, it doesn't fade away quietly. It explodes. Stars that are more than eight times as massive as our Sun become unstable. As the star's nuclear fires begin to wane, the star begins to implode under its own weight. But the collapse of the star rekindles a last burst of nuclear fusion. In less than a second, the star explodes in a tremendous burst of energy. The catastrophe spews the elements from its interior into space, like a puffball spreading its seeds across the landscape.

Eventually, these chemical elements will be incorporated into newly forming stars, and they will also combine and coalesce into planets. The process continues as stars die and new stars form. Our own Sun is built from the remains of stars that have come before it. And the planet on which we live, and the material of which we are made, contains the chemical elements from previous stars that have lived and died.

We are very much star stuff.

The explosion of a star has another remarkable outcome. It creates additional elements that could not be forged in the star's interior, either because the interior was not hot enough or because the elements themselves are inherently unstable.

For example, the explosion produces cobalt, whose brilliant blue compounds color Renoir's oil paintings. And it produces cadmium, whose brilliant yellow compounds keep Van Gogh's sunflowers sunny. It also produces uranium; mixed reviews there.

And the explosion produces platinum, gold, and bromine—three elements that sit quietly on the periodic table of the elements but have spoken loudly in human history through their connection with greed, envy, and corruption. Platinum and gold we can understand. But bromine?

Yes, bromine. I mention it because the story of nature is in part the story of us, so it helps to fast-forward every once in a while to remind ourselves of that. I'm thinking of the incident, a very long time ago, in which a tiny sea snail in the Mediterranean learned, through no fault of its own, how to attach two bromine atoms to a molecule of indigo. That innovation created a lovely purple hue. I have no idea what role the purple hue played in the snail's life but, as fate would have it, the Phoenicians and Romans learned to extract the material from the snail. They called it Tyrian purple, or royal purple. (In fact, *Phoenicia* means "land of purple.") The adjective *royal* gives a hint that the color was the province of the rich and powerful, a restriction whose violators often faced death.

The Roman senator Cato the Elder spoke passionately in defense of Rome's Oppian law, which absolved wealthy men from having to give their wives expensive gifts of "gold and purple."[2] The women rebelled, arguing loudly in the streets. How would anyone know that they were well off if they could not "make a figure, distinguished with gold and purple?" "Why," they wanted to know, "is the poverty of others concealed under this cover of a law?"[3]

For his part, Cato argued that "[if women are] to be set on an equal footing with yourselves, can you imagine that they will be any longer tolerable? Suffer them once to arrive at an equality with you, and they will from that moment become your superiors."[4]

Eventually, royal purple and indigo blue were no longer the sole province of the wealthy. The ancient Hebrews were instructed to include a single blue thread in their prayer shawl; it is thought to have signified that each person is royalty in the eyes of God. In fact, indigo became the people's choice. It dyed the fisherman's garments in Genoa (hence *jeans*) and the garments in Nîmes (hence *denim*), and it now gives color to more clothing than any other dye. It has been the most popular color every year for at least six thousand years.

The Acme of Elements

Of all the elements that a star creates, surely the greatest achievement is carbon. In its rough form, carbon is merely the charcoal in your backyard grill or the coal in a furnace or the graphite in a "lead" pencil. In its pure crystalline form, it is diamond. But in its most precious form, carbon is the basis for all life on Earth.

What makes carbon great is the company it keeps. It combines with other elements to form an astonishing variety of molecules. Elements such as hydrogen, oxygen, nitrogen, and sulfur are eager to partner with carbon. Carbon is the master of a thousand faces. It forms more kinds of compounds than perhaps all the other elements combined—an essentially unlimited array that makes possible the mind-boggling diversity of life.

The aroma of coffee, the flavor of mint, the sheen on a butterfly's wing, the nerve cells that carry your every thought—all are built from carbon compounds. So are the organisms that have yet to appear on our planet. So are the pharmaceuticals that have yet to be invented, and so are the compounds whose properties we can't even imagine at present.

I am always surprised when people ask whether there could be life based on elements other than carbon. Perhaps there could be. But the question usually sounds as though carbon and life on our planet were not exotic enough. My first thought is Samuel Johnson's old saying that "when you are tired of London, you are tired of life." When you are tired of carbon, you are tired of life. If you think life on Earth is pedestrian, you will want to see more of what nature has already conjured up on our planet. And carbon hasn't even *begun* to strut its stuff yet.

It almost didn't turn out that way. In fact, carbon owes its very existence to yet another of those miraculous features of nature's building plan.

You might think that fusing nuclei together to make heavier and heavier elements would be a simple process, but it is not. It is not like snapping together building blocks in a toy set. In fact, the nuclear reactions inside a star are highly complex. They are sensitive to the temperature and pressure inside the star, the number and kinds of elements already present, and the stability of the nuclei themselves. In turn, the stability of the nucleus depends on the energy levels that nature makes available to it.

And when it came to the element carbon, there appeared to be a problem.

By the 1950s, scientists realized that the chain connecting the elements had a weak link at the very beginning. At first, two helium nuclei fuse to form beryllium. Then, in principle, another helium nucleus would fuse with beryllium to form carbon.

But the beryllium appeared to produce a bottleneck, because it is extremely unstable; it breaks apart in far less than a trillionth of a second. In 1953, the British astrophysicist Fred Hoyle realized that the reaction could proceed if three helium nuclei collided almost simultaneously—provided that the carbon nucleus had a favorable energy level called a resonance. Hoyle was able to calculate what that energy level should be.

Did carbon actually have this predicted energy level? A team of nuclear physicists at the Kellogg Radiation Laboratory at the

California Institute of Technology searched for and found precisely the energy level that Hoyle had predicted.

Why should nature arrange for carbon to have just the right nuclear structure to ensure its formation? Hoyle put it this way: "A common sense interpretation of the facts," he wrote in 1982, "suggests that a superintellect has monkeyed with physics, as well as chemistry and biology, and that there are no blind forces worth speaking about in nature."[5]

You might argue that it is just serendipity. A lucky coincidence. But no, that can't explain it. A coincidence is when two old friends just happen to meet on a street corner after not seeing each other for years. They might just as easily have not met at all. If one had stopped for lunch, if the other had walked a little faster, if the light had not turned green when it did, then the outcome might well have been different.

But the creation of carbon is dictated by the laws of nature. Every carbon atom in the universe has the same structure. And every star with the same mass and history will produce carbon in much the same way. The universe's building plan—including the strength of gravity, the strength of the electromagnetic force, the strength of the nuclear forces, and the masses of basic particles guarantees the birth of carbon.[6]

In fact, it is not only the creation of carbon that is remarkable. If you step back and look at the palette of elements that nature creates, it is truly marvelous that they are created in the abundances needed to form planets and to nurture life. It didn't have to be that way. If the nuclear reactions all went to completion, all the hydrogen would be used up. All the carbon would be used up. The end product would have been a universe essentially filled with iron, the most stable element. That would be good for kitchen-ware, but not for life.

"Small, but Oh My!"

The weak nuclear force struts its stuff on the cosmic stage when a star explodes. The explosion itself is remarkable, because it is driven by what is ordinarily the most timid of particles—the neutrino. These particles are created in the interior of the Sun and every other star, where they carry off some of the energy from the nuclear reactions. A neutrino feels *only* the weak force. (It is similar to an electron, but it has no electric charge.) Right this moment, a trillion neutrinos are passing through you every second, without disturbing you in the slightest. Since they interact with matter only via the "weak" nuclear force, they pass through the Sun, through you, and through the entire Earth without a second thought. They can continue sailing on through the universe for billions of years.

But at the end of a star's life, when the star begins to collapse on itself, neutrinos are produced in such vast quantities that they deposit their energy with the surrounding matter in the star, generating a titanic explosion that sends the star's core flying outward at more than twenty million miles per hour!

Why does the neutrino exist? This is not a typical scientific question. But if we ask, In what way does the neutrino fit into the story of the universe? then surely its role in the explosion of the star would be the highlight of its existence. In the one second it takes a star to explode, the neutrino proves its usefulness to the cosmos, and to life everywhere.

Scientists have been able to simulate the explosion of a star on a supercomputer, though it is not easy. There are computer models in which the explosion fizzles, for one reason or another. Fortunately, nature knows how to accomplish the task, at least in enough stars to make a difference to the ongoing creation of the universe.

To summarize: A large star is not merely the workplace for creating the elements. It also has a mechanism to deliver these elements to the cosmos, a mechanism built into its very existence by the laws of nature. The mechanism is its own death.

Small Is Beautiful

If that were the whole story, we wouldn't be here. A massive star's explosion is so powerful that it annihilates anything nearby, including the star's own planets. So if there were nothing but large stars in our universe, no one would be around to tell the tale. Big stars live fast and die young, and anything around them will die young too.

There would be no apple pie baking in the oven.

Fortunately for all concerned, nature also creates long-lived stars. Smaller, gentler stars. Stars that welcome life.

Our own Sun is one of them. It will live a thousand times longer than the massive stars described earlier. Billions, not millions, of years. Like all stars, the Sun is busy creating chemical elements, but it will never get much past carbon, nitrogen, and oxygen. And it will never get credit for what it does produce, because it will not explode. The fruits of its labor will largely remain locked inside. If all stars were like our Sun, there might be no life anywhere.

But that's not what the Sun is all about. Nature has given it a role that is every bit as amazing as what the massive stars do. Our Sun is truly one of nature's wonders.

The Sun is an incubator. It provides the warmth and the light needed to nurture the development of life on Earth. It provides these virtually without change over the billions of years needed for life to evolve. The life span of the Sun is well matched to the time required for evolution.

Last spring I watched a robin build its nest and incubate its colorful blue eggs just outside my office window. As a chill New England rain began to fall, the robin spread its wings over its hatchlings to protect them from the rain. It was a striking gesture: When an animal is cold, it normally pulls its limbs in and curls into a ball to reduce its surface area and conserve its own warmth. But this robin, miserable and determined, sacrificed its own warmth to preserve that of its offspring.

It's one thing to spread your wings for a few hours a day. But the Sun has kept life on our planet comfortably warm for billions of years. This giant ball of unassuming hydrogen gas is perfectly constructed to do just what the robin does: keep life at a constant temperature. And it has no other way to do this than by doing what the laws of nature tell it to do.

Of course, if you happen to be on the planet Mercury, close in, the Sun will be happy to fry you. If you are on Pluto, prepare to freeze solid. The Sun couldn't care less. But here on planet Earth, the Sun is life's guardian. Every second of every day, for billions of years, it has provided a virtually constant warmth—and just the right warmth. There is a reason that ancient people worshiped the Sun.

This is the way nature works. Silently, of its own accord, of its own necessity. Again, it would not have been this way were it not for nature's building plan. You could easily imagine that the Sun's birth might have also been its death. That the moment its nuclear fire ignited, the whole star might have exploded. Or what if the Sun could not sustain its nuclear fires and fizzled upon birth? But it didn't. The Sun has been a faithful incubator for billions of years. It works.

In 1952, scientists were able to duplicate the Sun's nuclear fusion when they set off the first hydrogen bomb. But sustaining a fusion reaction has proved much more difficult. The industrialized nations have poured billions of dollars into the most sophisticated technology, with the goal of sustaining a fusion reaction in the laboratory. (One approach uses the world's most powerful lasers to compress and confine the nuclear fuel. Another uses powerful magnets to confine fuel that is too hot to be contained in any solid vessel.) A global consortium of nations has collaborated for thirty-five years to build a new facility that may create the first sustained fusion reaction; the project is called ITER (from the Latin for "the way").

To date, a fusion reaction has been sustained for the tiniest fraction of a second. Yet nature accomplishes fusion, sustained and steady for billions of years, with no technology at all. Take a large enough cloud of hydrogen, let it collapse under its own gravity, and the nuclear fires ignite and burn steadily for billions of years. Surely the Sun is nature's simplest and grandest machine. It is a machine built from the simplest of elements and constructed with no other instructions than the simplest of nature's laws.

The Giant Thermos

The Sun has another remarkable property needed for life. It is among the best thermos bottles in the universe.

At first glance, the Sun seems extremely hot and bright. After all, its surface temperature is nearly ten thousand degrees Fahrenheit, much hotter than the hottest welding torch or pottery kiln. And it is so bright that it would blind you were you to gaze at it. But when compared with the Sun's interior, the surface is positively cool and dark.

The Sun's interior is a thousand times hotter than its surface. And the interior is one *trillion* times brighter than the surface. If you were to peel away the Sun's outer layer, like the peel of an orange, and expose the central core, even for a moment, the heat and light released would vaporize Earth instantly. Fortunately, the tremendous heat from the Sun's nuclear fires is contained safely within. So not only does the Sun provide this constant heat and light over billions of years, it does so safely.

The Sun's surface is not merely a safe temperature. It is the *right* temperature to provide a spectrum of light that will drive the development of life on Earth. The light leaving the Sun has just the right span of energies to stimulate chemical reactions here on Earth.

Among the chemical reactions that light makes possible are the ones that give us sight. Were it not for sunlight, animals would have evolved to be blind. Of course, we see far more than sunlight. Thanks to our sight and our curiosity, we can detect light from the very edges of observable space, allowing us to sketch the portrait of a universe that we cannot hope to touch directly.

10

Into the Light

If the universe kept a diary, it might well have made a journal entry on one joyous day nearly four billion years ago, when the first life emerged on Earth. Whether that life developed in a warm pond, as Charles Darwin imagined, or in a deep-sea vent or other locale is not central to our story. And whether that life was driven by a chemical source of energy or some other source is also a detail.

On that special day, nature embarked on the greatest phenomenon and grandest adventure known: a self-sustaining set of chemical reactions that have grown and mutated and become extraordinarily complex but that have never come to an end. Every living thing has budded from that continuous, connected chemistry—from that tree of life. If only we could remember our own gestation and birth, we might better see our connectedness to the living whole.

Life poses a big problem for our story. Up to this point, we could clearly see how the universe followed inexorably from its building plan. Space, stars, planets—they all arise from the machinery of the universe. We can observe, we can simulate, we can explain the basics. But life is still shrouded in mystery. It seems paradoxical that we can dissect every life-form right down to the molecular level—there is virtually no aspect of any creature that we cannot probe in as much detail as we like—and yet life itself remains bookended by two of the greatest puzzles: at one end, how life originated, and at the other end, how it is possible for mere molecules to assemble themselves to be aware, sensate, conscious.

Let's look at a few of the universe's journal entries to get a sense of the challenges that life had to face, and to see how life innovated. Since we'd like to arrive at an apple pie, we'll focus on plants—and what plants have done with two of the simplest molecules in nature: carbon dioxide and water.

Harnessing the Light

The day I have in mind is the one in which a living cell was first able to harness the energy of sunshine, in the process we now call photosynthesis. Sunshine is produced by the fourth great force of nature: electromagnetism. In fact, light is the carrier of this force, transporting it the huge distance from the Sun to Earth. It is also the force responsible for holding atoms and molecules together and for guiding their chemical reactions. Electromagnetism drives the chemistry of life. It is like an artist's small sable brush, responsible for crafting the finest, most intricate details of our world.

At first glance, photosynthesis might seem like a simple process: the raw materials are carbon dioxide and water, and they react with each other using the energy in sunlight. The end products are oxygen and sugars, which are then used to build up the molecules of life. We don't have all the details of the original process, but today, the process is complex.

The chlorophyll molecule—the poet Walt Whitman's "hopeful green stuff"—serves as the antenna that captures the energy of sunlight; it passes this energy to a factory of one hundred proteins that use the energy to transform carbon dioxide and water into the more complex compounds of carbon. When the chlorophyll molecule was first synthesized by humans in 1960, the feat required the world's greatest organic chemist, Robert Woodward; a dedicated team of seventeen postdoctoral researchers; fifty-one separate steps of chemical synthesis; and several years to accomplish. But a living cell synthesizes the same chlorophyll in less time than it takes to read the morning newspaper. And it was able to do so more than a billion years ago. Clearly, life is among the greatest of chemists.

Photosynthesis is one of the longest-running shows on Earth, and with good reason.[1] Virtually every living thing on Earth depends on the ability of plants to capture the energy of sunlight, store it in carbon compounds, and cycle it through the community of life.

Every blink of an eye, every trill of a bird song, every fiber in a spider's web ultimately depends on the quiet work of plants, harvesting the sunshine. Our thoughts, dreams, and memories are all powered by it, all woven from the electrical energy captured from the Sun and passed from molecule to molecule. As the biochemist Albert Szent-Györgyi summarized it, life is driven by "a little electric current, kept up by the sunshine."[2]

In autumn, when a leaf's work is done and it prepares to fall from the tree, even that simple act requires energy. (A special enzyme must dissolve a leaf's woody connection to its twig, allowing gravity to take over. If a tree were to die before this happens, the leaves would remain attached.)

Ultimately, photosynthesis powers not only life but also our *way of life*. There is an apple pie baking in the oven now, and the oven itself owes its existence to the fruits of photosynthesis. The oven's steel and aluminum were coaxed from their ores using furnaces stoked by coal, which is the remains of plants long gone: forests that lived and died millions of years ago. Even the oven's plastic dials are the work of plants; they are synthesized from petroleum, which is also the remains of organisms whose high-energy molecules have been transformed into fuel over millions of years. The electricity that runs the oven is generated by oil and natural gas as well.

In fact, all the world's cities, with their walls of glass and steel and their sidewalks and subways, are dependent on the busy work of plants, in the form of fossil fuels. The entire human-made world is ultimately founded on the labor of the sun harvesters.

Take away photosynthesis, and the world would be as barren as the Moon.

I imagine that you are not impressed. And why should you be? If you have pulled weeds from your garden, you know that they seem to come back as soon as you turn your back. You can bring a canteloupe home from the store, and in a flash a mold devours it. Life seems to pop up without effort. Green is everywhere. We take life for granted.

So I'd like you to try a simple experiment that might change your mind. Take a bottle of ordinary seltzer—a mixture of carbon dioxide and water. Make sure it is in a *very* stable container. Place it in your refrigerator. Now leave it there for one hundred million years. We'll come back to the seltzer in a little while to see how it is doing.

The Dark Side of Light

Our next peek at the universe's diary takes us to an especially critical day several billion years ago. The invention of photosynthesis had created a crisis that could have destroyed life on Earth, had nature not been able to innovate. The problem was with one of the end products of photosynthesis: oxygen. Over many millions of years, the gas started to accumulate in the atmosphere. At first, much of the oxygen reacted with iron in the Earth's surface, rusting it. But eventually the levels of oxygen started to rise significantly.

You might think that you could never have enough oxygen, but in fact the substance is remarkably toxic. Oxygen is so reactive that it will gradually combine with and destroy most compounds that are not already fully oxidized. Iron rusts. Paper yellows and browns with age, as though it were slowly burning. Butter turns rancid.

We would turn rancid too, in quick order. But fortunately, at some time in the deep past, our ancestral cells figured out how to deal with oxygen: sequester it in the cell and then quickly use it to burn food. This not only gets rid of the oxygen, it provides energy needed to power the cells' infrastructure. Today, more than 90 percent of the oxygen in our cells is used by one enzyme, cytochrome oxidase, which is common to virtually all living things.

Despite the safeguards within cells, our bodies still cannot handle excessive oxygen. For example, in the fifteen years between 1942 and 1957, almost ten thousand premature babies were rendered blind by high-oxygen therapy.[3] Oxygen is now diluted with other gases before being administered.

Even though oxygen posed an ever-present threat to life, it wound up providing two miraculous benefits that enabled life to flourish on land. First, cells that metabolize their food with oxygen produce almost twenty times more usable energy than cells that don't use oxygen. This extra energy allowed animals to grow larger and swifter. (Once you leave the buoyancy of the ocean, you need more energy to support your weight.)

Second, oxygen forms a protective layer of ozone (O_3) in the upper atmosphere. The ozone layer absorbs much of the Sun's damaging ultraviolet light, which would otherwise destroy the DNA in the cells of land animals. Without oxygen in the atmosphere, we might still be a single cell at the bottom of the ocean.

Variations on a Theme

As we flip through the universe's diary, we see so many proud entries; nature seems to be the master innovator. Here's an example: All higher plants must open their pores to take in carbon dioxide. Over the past three hundred million years, as carbon dioxide levels have gone up and down, plants have learned how to adjust the number of pores in their leaves accordingly, so that they take in no more carbon dioxide than they can use. This also helps the plant conserve its water, some of which evaporates through the pores.[4]

The cactus on my windowsill comes from an especially clever lineage that has taken this one step further. In the desert, during the day, a cactus doesn't dare open its pores at all. If it did, its precious water would evaporate and the plant would become a shriveled mess.

In response to this challenge, desert cactuses have become the Dunkin' Donuts of the cactus world: they make their sugars at night. The cactus has evolved a new step in photosynthesis that allows it to capture the energy of sunlight during the day and store it temporarily, without ever opening its pores. In the cool of the night air, the cactus opens its pores wide, drinks in the carbon dioxide, and then uses the energy it captured during the day to manufacture its sugars.

A sunflower plant is equally clever. It has developed a mechanism to turn its growing tip so that it always faces the Sun during the day, the better to capture the Sun's light. The plant starts out facing east in the morning and winds up facing west by the evening. Overnight, it returns to face east. This routine helps the plant grow faster.

However, once the flower has appeared, the plant no longer tracks the Sun but instead keeps its flower pointing east. Experiments have shown that by capturing the morning sunlight, the flower warms up quickly and attracts more early-rising insect pollinators. After its flowers have formed, the plant is only interested in producing seeds, so it needn't expend energy to turn with the Sun and grow faster. Plants are not dummies.

In fact, plants are much smarter than we give them credit for. As spring arrives, many plants test the season by checking whether the days are getting longer, the temperature is rising, and so forth. Remarkably, some plants go further; they double-check their assessment by actually measuring the color of sunlight. As the Sun gets higher in the sky, its light becomes slightly bluer. As it gets lower in the sky, its light is yellower and, by the time it sets, reddish. Plants are able to sense how high the Sun gets just by testing the color of sunlight.

No wonder many animals have developed such a close association with plants. When you think of a tropical coral reef, you think of fish. But the entire community of life there rests on the work of plants. Corals can inhabit nutrient-poor tropical waters thanks to their partnership with tiny algae. The coral polyp provides the carbonate structure that will ultimately become a coral reef; and the algae provide the food, some 80 percent of which goes to the coral. Without tiny algae to support the larger reef community, the fish and other living things would vanish.

Or consider lichens, which have mastered a partnership between fungi and algae, allowing both to grow on bare rock. The fungus provides the acids that slowly dissolve the rock; the algae provide the carbohydrates. Thanks to this partnership, it is reported that lichens cover up to 8 percent of Earth's land surface!

Even plants partner with plants. I'm thinking of the ghostly white Indian pipe, *Monotropa uniflora*, which I spotted growing in Acadia National Park one summer. It has lost all of its chlorophyll and no longer bothers making its own food. Instead, it taps into the fungi in the soil, which in turn tap into the roots of the surrounding forest trees. The trees, of course, do make food. The Indian pipe can be forgiven for being a parasite: thanks to the trees, so little sunlight reaches the forest floor that it seems reasonable to survive by attaching oneself to the big players in the forest. A little sunshine goes a long way through the food chain.

Living Far from Equilibrium

Light and life are so familiar to us that we take them as a given. So it is worth pausing for a moment to consider just how extraordinary and unexpected it is that life has been achieved by mere molecules.

You can see the challenge that life faces by contrasting a child's chemistry set with a morning glory seed. A chemistry set is full of promise at the very beginning, resplendent with colorful liquids and solids and experiments that produce dramatic color changes and crystals and bubbling reactions. But, as every child knows, when the chemical reactions have run their course, the show is over. At that point, there is only a mess to clean up. Slime, gunk, test tubes that can never be cleaned. The mixture of chemicals has reached *chemical equilibrium,* the state at which no further change happens.

Essentially all of the chemical reactions that we learn about in school are of this type. The reaction goes to completion, something happens, you get a product, end of story.

By contrast, the morning glory seed is dark and unpromising to begin with and does nothing for a long while. A child needs patience to watch it germinate and grow. But it unfolds in time *without ever coming to an end.* The flowers produce new seeds, which, in turn, produce new flowers and so forth, so long as the raw materials for new plants are available.

Life lives far from chemical equilibrium. Not for a single second has this growing, spreading phenomenon reached chemical equilibrium in three billion years. Of course, individual organisms die, fall from the tree of life, and return to the soil or sea. They return to the world in the form of water and carbon dioxide and other simple molecules whence they came. But the community of life endures, and its chemical reactions continue far from equilibrium: countless molecules mandated to create life just by obeying their own nature. And all of it driven by the energy of sunshine.

To live far from equilibrium is to continually struggle against coming to equilibrium—that is, eternal rest. Death.

I like the way the naturalist Joseph Wood Krutch put it. He wondered why we have so much more affinity for a living fern than we do for a frost flower—the fernlike pattern of ice crystals that forms on window panes in winter. "There is something about the fern leaf," he wrote, "which stirs us . . . as the frost flower cannot. We know that the fern is like us, while the frost flower is not. In some sense all living things are allied in some sort of struggle against all that are not living."[5]

This alliance may explain why Joseph Banks, the botanist on Captain James Cook's voyage whom we met earlier, gave his wife a piece of dried moss to wear as a brooch, in place of the diamonds she really wanted. As historian Andrea Wulf tells it, "While he thought it a gorgeous botanical specimen, his wife found it boring and unsightly. When she refused to pin it to her blouse, Banks called her a 'Fool that She Likes diamonds better, and Cannot be persuaded to wear it as a botanist's wife Certainly ought to do.'"[6]

Galileo would likely have approved of Banks's gift. "What could be more foolish," he wrote, "than to say that gold, silver, and gems are precious, and that earth and mud are merely vile?"[7] If earth were as scarce as jewels, he continued, then every prince would "give a sackful of diamonds and rubies . . . simply to have enough earth to plant a jasmine in a little pot, or to grow an orange tree from seed."

To Galileo, the very permanence of a diamond was its greatest flaw. It was the changeability of life that gave it meaning, even with its inevitable outcome. "I think," he wrote, "that those who exalt incorruptibility, immutability, etc., so highly say these things because they are so anxious to have a long life and are so terrified of death; they don't realize that if men were immortal they would never have been brought into the world. It would serve them right if they were to encounter a Gorgon's head which would turn them into jasper or diamond statues, so that they could become more perfect than they are."[8]

The Story Heats Up

Ironically, photosynthesis has led the world to the brink of catastrophe. That ubiquitous, silent process on which all life depends has come back to haunt us. And this time the difficult problem that nature faces is *us*. I bring up this episode only to help place our species where it might belong on the arc of nature's story.

The makings of the catastrophe began about 360 million years ago, at the start of a period that geologists call the Carboniferous. For the next sixty million years, trees began to spread over Earth. They were very different from the trees of today: not yet well rooted and much higher in lignin and other tough, carbon-based fibers. When these trees fell, they were slow to decompose; the microbes to help with the decomposition had not yet evolved. As a result, the carbon in these trees was not recycled into the atmosphere to be taken up again by new trees. Instead, the carbon was buried deep underground and ultimately converted through heat and pressure into coal. About 90 percent of the world's coal beds were formed during this period.[9]

Some three hundred million years later, our species began to dig up the coal beds. Ever since the Industrial Revolution began in the eighteenth century, we have been burning increasingly more coal, as well as oil and gas, which also are the remains of photosynthesizing organisms. We have put a million years' worth of carbon dioxide into the atmosphere in the space of just a few years, on a grand scale. Of the ten largest companies in the world by revenue, more than half are now fossil fuel energy companies.[10]

Carbon dioxide is not only a raw material for life. It also helps to regulate our planet's climate by trapping some of the heat that Earth would normally radiate back into space. Earth's habitability depends on a remarkable, complex cycle that removes carbon dioxide from the air and then eventually returns it. The rates of removal and return must be in balance for Earth to remain habitable. While Earth's climate has fluctuated markedly over history, the natural cycles have kept the fluctuations within limits. (Our neighboring planets, Mars and Venus, were not so fortunate.)

The universe's journal entry for this episode is blank because the story is still taking place. But our own species has been keeping a diary of sorts. Three of those historical pages will serve to introduce nature's most unusual innovation.

The first entry is by the Swedish physical chemist Svante Arrhenius, who calculated that rising levels of carbon dioxide from the Industrial Revolution would gradually cause Earth to warm. In his 1908 book, *Worlds in the Making: The Evolution of the Universe,* he found the prospect of global warming something to be welcomed. "We often hear lamentations," he wrote, "that the coal stored up in the earth is wasted by the present generation without any thought of the future . . . [but] we may find a kind of consolation in the consideration that here, as in every other case, there is good mixed with the evil." As the level of atmospheric carbon dioxide continued to rise, he predicted, "we may hope to enjoy ages with more equable and better climates, especially as regards the colder regions of the earth, ages when the earth will bring forth much more abundant crops than at present, for the benefit of rapidly propagating mankind."[11]

For our second journal entry, fast-forward about fifty years to a 1965 report called *Restoring the Quality of Our Environment.* The report was issued by the environmental pollution panel organized by President Lyndon Johnson's Science Advisory Committee. Like Arrhenius's book a half century earlier, the report noted that rising emissions of carbon dioxide were gradually heating Earth. The report accurately predicted that carbon dioxide levels would continue to rise as industrial society continued to burn fossil fuels. (The actual rise of 33 percent by the year 2000 exceeded the report's prediction of 25 percent.) And the report recognized the possible effects of an increase in atmospheric carbon dioxide, beyond global warming. Among them were the melting of the Antarctic ice cap, a rising sea level, and the warming of the oceans. These predictions are being borne out. The report concluded, tersely, that the climate changes induced by increasing levels of carbon dioxide "could be deleterious from the point of view of human beings."[12]

The report had only two recommendations, both of them involving large-scale tinkering with Earth's surface. One was to reflect sunlight back into space "by spreading small reflecting particles over large oceanic areas"—namely, five million square miles of ocean. The other was to increase the cloudiness of Earth by seeding the formation of high-altitude cirrus clouds in an attempt to modify Earth's atmospheric circulation. The report made no mention of ameliorating climate change at its source—for example, by mandating reductions in fossil fuel emissions, or by researching and supporting alternative energy sources, or by defining and supporting a national program to conserve energy.

Fast-forward another fifty years. In a report to Congress dated July 23, 2015, the Department of Defense noted that "the National Security Strategy, issued in February 2015, is clear that climate change is an urgent and growing threat to our national security, contributing to increased natural disasters, refugee flows, and conflicts over basic resources such as food and water. These impacts are already occurring, and the scope, scale, and intensity of these impacts are projected to increase over time."[13]

What an interesting twist to the story of photosynthesis. Nature has come up with an innovation—us—that can now harness the energy of the Sun directly, using solar cells, artificial photosynthesis, and other new processes. But we haven't quite managed our community affairs yet. Though the verdict on our species is still out, our emergence heralds the most revolutionary stage in the universe's development.

The Creation of Creativity

Most scientists I know are hesitant to point to humans as the pinnacle of nature's achievements; they are too aware that nature has created a problem child. Nevertheless, with us, the universe has reached a remarkable milestone: the creation of creativity itself.

For some reason, people feel compelled to invoke Beethoven's symphonies as the exemplar of creativity. Is that really necessary? We are creative every time we decide what to wear in the morning, what to serve for dinner, whom to invite, how to improve a recipe, what to plant in the garden, what to write to our friends. We are creative when we plan a vacation or patch a leaky faucet before the plumber can do the job right. We are creative when we lie to others—and even more creative when we choose an ethical life. We are creative when we dream. Are we not all about creativity?

And surely our species is not alone in this talent. I once fell in love with a friend's dog, and I think the feeling was mutual: every time I visited, Yo-Yo would grab an old shoe, circle two or three times with it, and then deposit it at my feet. I knew that dogs like old shoes, and that they do bear gifts, but I didn't understand the circling part. What was that about?

Years later, I realized that the circling was Yo-Yo's version of gift wrap. Why, after all, do we wrap a gift with colorful and expensive paper, only to have someone rip it off and throw it away? We do it to show that the book inside is not just any old book (or any old old shoe) but is special. It's for you. We do it to delay the gift a little, to add suspense. And we do it to say, This is from *me*. I'm the one giving you this present. Yo-Yo was creative with her version of gift wrap.

With the emergence of intelligence and creativity, it is as though nature has created a brand new force—one that controls the other forces of nature and uses them to its own purpose. Our species has started to edge out nature, as though we had taken over the mantle of creation—and destruction as well. For better or worse, you and I are not bound by a building plan; we have free will. (I once asked the molecular biologist John Cairns whether he believed in free will, a question that I regretted as soon as I uttered it. Without hesitation, he replied, "Do I have a choice?")

To dwell on our species would be a diversion from our story; it is a different conversation. But I like the way William Shakespeare handled the interface between us and the rest of nature in *The Winter's Tale*. Perdita explains why she lets her garden go barren rather than grow variegated strains of carnations created by the plant breeders. These flowers, she says, are "nature's bastards," and she notes that the art of plant breeding cannot compete with "great creating nature."[14] At first, Polixenes points out that the plant breeder's art "does mend nature"—that is, improve on it. But then he corrects himself and says, "change it rather." After all, who is to say that our creations outshine nature's or vice versa? And finally he concludes, "But the art itself *is* nature" (emphasis added). For better or worse, we are part of nature, and everything we do might be seen as simply nature's course.

A Grand Cycle

The apple tree in my backyard is on its last limbs, having weathered one too many New England winters. But the tree still manages to carry out the ritual of photosynthesis that its ancestors have faithfully observed every day of every spring, summer, and fall, year in and year out, good times and bad, for three billion years.[15]

A limb from the tree has fallen to the ground. The upward optimism of electromagnetism has lost out to gravity. But new limbs will grow, as they have year after year, with the help of a little sunshine, our star reaching across space to power life.

Time to return to your refrigerator and look at the bottle of seltzer. It is still just a mixture of water and carbon dioxide. And in a hundred million years, it will still be just a mixture of water and carbon dioxide. It will have accomplished nothing. It is at chemical equilibrium—eternal rest.

But in that same time, the world's plants and animals will have cycled carbon dioxide and water in and out of their living bodies hundreds of millions of times. They will have sustained themselves from generation to generation. They will have built entirely new forms of life, some of which will have transformed the planet. Thanks to the Sun's light and to the molecules of life, the community of living things is able to survive far from equilibrium.

The bottle of seltzer has no story to tell, yet the universe does.

The apple pie is ready to come out of the oven. We didn't really create the pie from scratch, for as Carl Sagan said, we would first have had to create the universe. And yet we inhabit this seed of a universe and are present at its unfolding. And whether or not the universe had us or the pie in mind when it started out, now, at this moment, we are all a fruit of its labor.

Part Three

IS LIFE MERELY A ROLL OF THE COSMIC DICE?

11

The Great Inventor

Imagine a high-tech fabric that is soft and flexible, yet so durable that it comes with a lifetime warranty. As you grow, it stretches with you. It keeps you warm in winter and cool in summer. If you spend time in the sun, it won't fade; in fact, it darkens to protect you from the Sun's harmful rays. Should you stain it, you needn't worry: it sheds stains and is self-renewing. Should you tear or scorch it, you needn't reach for a sewing kit: within just a few days, the fabric mends itself.

You are wearing that fabric right now, of course: it's your skin. If you happen to be a certain desert cactus, your skin produces fine white hairs that reflect the sunlight and keep you cool. If you happen to be a polar bear, your white hairs are hollow, the better to insulate you and keep you warm. If you happen to be a baleen whale, your skin produces a mouthful of hair instead of teeth, the better to capture the microscopic food you eat.[1] If you are a cuttlefish, you can change the color of your skin at will.

Life is the greatest of all inventors. It is so varied, so beautiful, and so complex that one has to ask, How could such an intricate tapestry possibly have been woven? It is difficult to imagine that a polar bear's fur could have been written into the universe's building plan. What would that even mean? On the other hand, how could it have arisen by sheer chance? What does *that* mean? Is the polar bear some combination of both—the luck of the draw *and* the consequence of nature's building plan?

Then too, we know that the polar bear's fur reflects its environment. A polar bear is white because ice and snow are white; the camouflage helps the bear escape detection by its prey and therefore survive. On the other hand, how did nature conveniently come up with white fur? There has to have been an evolutionary pathway to get there, and molecules had to be obligingly recruited to actually produce the white fur. What a strange partnership between building plan and environment!

Certainly, the preconditions needed for life to emerge are written into the building plan, as we have seen. The mix of chemical elements, the vast assembly of rocky planets scattered throughout our galaxy, the stable host stars that warm these "home worlds" for billions of years and bathe them in the light needed to power the chemical reactions of life, the abundance of water—these and many other factors emerge naturally from the universe's initial conditions. And we know this because we can model the formation of stars, galaxies, and planets using nature's building plan as input. But when it comes to life, there is an abyss between the physical world and the living world. We cannot yet deduce that there must be life.

Nevertheless, there is substantial and growing evidence that allows us to draw tentative conclusions. The remainder of this book presents the chief lines of evidence that life really is written into the universe's building plan. Most of the evidence is scientific, but there is also circumstantial evidence to consider. So I will present the case not as it would appear in a scientific research article but rather as it might be presented in a trial; we will thus have to use our judgment as we sift and weigh the evidence.

In that spirit, I'll summarize the case for you as the universe's lawyer might.

"Ladies and gentlemen of the jury, the only thing that my client, the universe, seeks is credit for life, the greatest of its creations, for which it has toiled tirelessly for billions of years. The other side has besmirched my client by carelessly tossing around terms like 'chance, pure chance,' and 'accident' and 'concatenation of improbabilities' and the like—creating an impression among the public that my client was somehow not involved in its own creations.

"We will see that my client *invented* randomness and wove it into its very rules—which are patented, by the way. The universe has used chance for its own creative purposes, to help create the vast diversity of life that graces the planet—without which, may I remind you, you wouldn't be here. My client has provided the raw materials on which chance works and, more importantly, provided the rules that determine how these materials behave: how they assemble, how they mutate, and ultimately how they give rise to life.

"Would you deny the great painter Jackson Pollock ownership of his creations simply because he dribbled paint on his canvas and didn't account for the location of every drop of paint? Of course not. My client is not claiming to have determined the shape and kind of every living thing. My client doesn't even know how you will rule in this case. That is the beauty of the universe's master-work: the variety and surprise of life. But my client invented the underlying machinery of life and endowed atoms and molecules with the properties that make life possible. It is a stupendous achievement.

"My client asks only that you hold your beliefs and conclusions in abeyance until we have had a chance to consider all the evidence. Thank you for your attention."

The Universe's Third Great Phase of Creation

In the remainder of this chapter, and for the rest of the book, I want to make plausible the idea that life arises naturally from the laws of nature. We'll examine many lines of evidence for this. Perhaps no one clue is conclusive in itself, yet collectively the clues form a prima facie case.

The evidence will support three claims:

1. Nature's crowning accomplishment is the creation of *the molecular machinery of life,* rather than any particular life-form.

2. The machinery of life is set up to preserve itself by replicating itself faithfully. To do this, it relies on the properties of molecules, which are ultimately dictated by the universe's building plan.

3. The machinery of life is also set up to innovate—to generate as many life-forms as possible, including intelligent life. The universe uses chance events to generate this diversity. But the viability of all the life-forms in existence is the result of molecular interactions whose properties, again, are dictated by the universe's building plan.

In this view, the machinery of life is the agent of creativity. It represents the universe's third great phase of creation. Evolution is not something that simply happens to life; change is guaranteed by the machinery itself.[2]

The machinery, or infrastructure, has several astounding properties. Before we consider them, let's remind ourselves what we mean by *the machinery of life*. What *is* life's infrastructure?

Life's Machinery: Ancient, but Not Primitive

The first living cells were not merely alive; they already contained the infrastructure for innovating, for creating new varieties of life. By *infrastructure*, I mean the molecular machinery that is common to all living cells and that supports the basic functions of life; we needn't define it more precisely than that. A portion of this machinery can be seen in Figure 11.1.

This photograph was made by taking a living cell and disrupting it with a single drop of dishwashing detergent. The fragments of the cell were then spread out on a slide and examined in an electron microscope, which uses a beam of electrons rather than light to create an image. These wonderful microscopes are housed in small, dark rooms in which you can spend many hours peering into the hidden world of the cell, a universe of its own.

As you examine the photo, keep in mind that you are not merely looking at the machinery of life; you are looking back in time as well. This infrastructure is common to all living things, as it has been for billions of years.

The thin strand running diagonally across the image is a tiny portion of the cell's DNA. Four regions along the DNA are actively at work. Each of these regions is a gene, which contains information needed to build the cell.

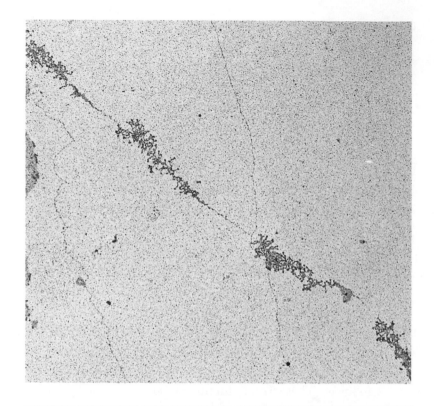

Figure 11.1. A strand of DNA runs diagonally across the picture. The four furry regions are genes, where the genetic information is being read and transcribed into RNA. This "machinery" has remained essentially the same for billions of years and is found in every living cell. Magnification approximately 20,000×. Credit: Photograph by the author.

By itself, the DNA would sit quietly and do nothing. But the information in the DNA is being "read" by hundreds of protein molecules that move along its length, from right to left in the photograph. The proteins are too small to see individually, but you can easily see the product of their work: the threads hanging from the four genes, looking like tiny Christmas trees. These are strands of RNA, which contain a copy of the information in the gene. The RNA carries this information to other parts of the cell.

Read is the right word for this process, because the information stored in DNA is a true language. The English language uses twenty-six letters, Greek uses twenty-four letters—and DNA uses only four letters. They are *A, C, G,* and *T*. These letters refer to the chemical subunits of DNA, which are bonded together and arranged one after the other, like the letters in this sentence. The subunits are too small to see even in this highly magnified image. Each gene in the picture happens to contain well over a thousand letters, including the sequence

   . . . GATGATCAGCCACACTGGGACTGAGAC . . .

The sequence of chemical letters looks like gibberish, just as the words in this book would look like gibberish if you didn't speak the language. But every living cell is able to interpret the information stored in DNA.

RNA molecules have many functions, but the primary one is to direct the synthesis of protein molecules. They do this at tiny molecular workbenches that are also made partly of RNA. In fact, the RNA strands that you see being made in Figure 11.1 are destined to become part of these workbenches, so they will help the cell manufacture more proteins.

Note that the spaces between the genes do not appear to be active. Nevertheless, they are essential. They contain the punctuation marks that help reader proteins know where to start and stop reading—just as capital letters, periods, and paragraphs help you to read this book.

These regions also control when and where the nearby gene becomes active. Just as a factory is dependent on all the employees knowing who does what and when they should do it, so is a living organism dependent on coordinating the switching on and off of genes. In fact, one of the big surprises of recent research is that human beings have far fewer genes than you might expect, given our complexity. It appears that the timing and control of our genes, rather than the sheer number of them, is key to building our complexity.

Even from this brief glimpse, you can get a sense of the complexity of life's machinery. But the infrastructure of life has several properties that are truly extraordinary.

Life's Machinery Is Extremely Robust

We think of life as the epitome of change. But the genetic apparatus you examined in Figure 11.1 was first invented billions of years ago, and it has remained virtually unchanged for all that time. This machinery of life is so robust as to strain credulity. If this image had been made a hundred million years ago, it would look much the same—and it could have been made using a cell from virtually any organism on Earth.[3]

This extreme stability of life's infrastructure suggests that, far from being an "accident," life is a necessary part of the universe.

In contrast to the creatures around us, which took billions of years to evolve, life's infrastructure emerged very quickly. The origin of life has been pushed ever further back in time as fossils bearing evidence of life continue to be unearthed. It now seems that life appeared almost immediately after the planet became habitable, where "immediately" means fast compared with the long time scales typical of geological processes.

Earth was formed about 4.5 billion years ago, but it remained uninhabitable for nearly the next half billion years. Asteroids and comets continually collided with the newborn planet, creating a molten surface that sterilized our world. This period in Earth's history is aptly called the Hadean period.

The oldest fossil evidence for life now extends back nearly four billion years to the ancient organisms known as blue-green algae, or cyanobacteria. The descendants of these pioneers can still be found growing in a few places, such as the west coast of Australia, where they aggregate into long, distinctive filaments that form layered mats as they die and are preserved. Although the interpretation of fossils is always open to question, fossils very similar to the present-day organisms have been uncovered in Australian rocks dating from nearly four billion years ago.

The machinery of life has survived unscathed for so long that it ranks as one of the universe's most enduring creations. You would have to struggle to find rocks that are three billion years old, yet something older than that—the machinery of life—is right under your nose, so to speak.

The "mere molecules" that form life's infrastructure kept watch over Earth long before our species fought its wars, long before the Pyramids were built and crumbled, long before the continents crashed into each other and drove up the mountain ranges, long before there was the sound of life on land.

Try comparing life's long trajectory with that of human culture. Some years ago, the Museum of Science in Boston hosted an exhibition of archeological finds from Caesaria, the ancient city on the Mediterranean that was built by King Herod. I was struck by a display of eleven ceramic oil lamps. Each lamp had the same basic design, including an opening in which to put the oil and a small hole for the wick. One or two of the lamps also had a handle for hanging the lamp on the wall.

I wondered whether all eleven lamps had been made by the same artisan or perhaps by a collective of artisans.

To my astonishment, it turned out that each lamp was from a different century. The lamps spanned more than one thousand years of history—yet the design had remained unchanged. Considering the pace of technological change today, it is hard to imagine that we would even recognize what form lamps will take eleven years from now, let alone eleven centuries.

And yet this constancy of design pales in comparison to nature's penchant for preserving designs that work. Almost immediately after it came into existence, life found mechanisms to preserve its most important molecules. It had mechanisms to repair DNA and proteins. We know this because we bear some of these ancient genes today.

We marveled that the Sun could provide a reliable, constant source of warmth and light for billions of years. And yet something as fragile as a gene, in something as vulnerable as a living thing, has also remained safe and sound and functioning and recognizable after a comparable span of time.

If you happen to encounter a machine or device that you think will still be useful four billion years from now, do not hesitate to invest in it! If you have a photograph, a song, or a favorite book that you think will still be readable four billion years from now, by all means save it in your attic. But be prepared for the continents to have drifted, to have been drawn again into the Earth and recycled, for the rocks you stand on today to be long gone.

The machinery of life not only has lasted that long but has spread itself over the planet, still churning out the latest renditions of life. One consequence of this conservation is that life is constructed from interchangeable parts. You can take a gene from a fish, put it in a tomato, and the gene will function beautifully, even though these species have been unrelated for hundreds of millions of years. (Try installing a part from last year's washing machine into this year's model and see what happens!)

This longevity of life's infrastructure, this robustness, is evidence that life is not an afterthought to the universe. It is not an epiphenomenon that accidentally showed up and then vanished. For some reason, and in some way that we don't yet understand, life is part and parcel of the universe.

Life Is Extremely Diverse

The machinery of life doesn't merely survive, it invents. It churns out an unending gallery of life. Midges, crocodiles, warblers, strange vines that don't yet have Latin names, whole ecosystems where you can be lost forever if you step off the trail. Tiny plankton that look as though an architect designed them. Creatures that live on your skin. And, of course, the countless, nameless beings that are now extinct and whose stories will never be known.

Life's robust molecular machinery, its infrastructure, created all of these. It has been spinning out life-forms with as much abandon as a sailor weaves tales of exotic lands. Do it, and never stop doing it. Life's infrastructure indeed relies on chance, among several other strategies, in order to continually expand and improve its product line. However, the viability of any particular life-form rests not on chance but on the properties of its molecules, which in turn derive from the universe's building plan.

How does nature achieve such diversity? In part, with proteins. They are the second essential part of life's infrastructure. DNA may be life's instruction manual, but without proteins, that manual would just sit there unread. Life is a universe in miniature, and proteins are what give it substance. We've already seen that proteins read a cell's DNA, determine which genes get turned on and off, and control the development of a living organism.

Proteins are also the architects and designers of nature's materials. The golden mane of a thoroughbred is made of protein. So is the silk of a spider's web, weight for weight stronger than steel. Proteins make up the bulk of our muscles—and if you've been bulking up, consider that plants invented muscle proteins first; they use them quietly inside their cells.

Proteins tell the sequoia tree how and where to lay down its giant pillar of wood; about eight hundred different genes are needed for this engineering feat. Proteins tell the butterfly how to craft the brilliant patterns of its exquisite wings. They instruct seashells how to lay down their exquisite patterns of mother-of-pearl—a construction so strong that scientists and engineers are trying to mimic it using synthetic materials.

Proteins are the master chemists of our world. The aroma of coffee, the flavor of vanilla, the scent of new-mown hay—all are the result of molecules that are crafted by proteins within a living cell. Most antibiotics are proteins, or are created by proteins. So are the antibodies that fend off the microbes in the world at large, every second of every day.

All this diversity arises from the utmost simplicity. Proteins are made from just twenty different building blocks: the amino acids, strung together in a chain. The sequence of building blocks is dictated by the cell's DNA.

The three-dimensional shape of a protein is what makes it useful to a cell. In principle, any given protein chain can fold into a vast number of shapes, just as a long length of yarn can fold into many different tangled messes. But in practice, nature selects for proteins that automatically fold into a single shape, step by step, as they are manufactured by the cell. An example is hemoglobin, the protein that carries oxygen in the bloodstream. You can unfold hemoglobin by heating it up; when it cools, it will refold into its original shape. Nature sometimes assists this process with so-called molecular chaperones, which are proteins that assist the folding of other proteins.

In fact, it is breathtaking that anything works at all within a cell, because all proteins are "sticky" to various degrees; the wonder is that the molecules don't all coagulate into a nonfunctional mess. Somehow, the thousands of different proteins that make possible life's diversity are all able to coexist.

Life Is Ubiquitous

I often hear people say that we can't really predict whether life is rare or common in the universe because, after all, we only have "an N of 1"—only one example of life: Earth's. That may be true, but we know much more about life than that it simply showed up once. The salient feature is not that there is life on Earth; it's that life is *ubiquitous* on Earth.

Life has swept over the planet for billions of years, diversifying into millions of species that now occupy virtually every imaginable part of the planet's surface. There are microbes that live in the deepest coal beds, and little worms that have been found in the cracks in rock almost a mile underground. There are microorganisms that live in the clouds, others that grow in the solid ice of Antarctica, and still others that thrive in the near-boiling-hot pools of Yellowstone National Park. And there are several pounds of microorganisms growing right now in and on you and me.

There are geese that fly over the Himalayas, where you and I would gasp for air. There are birds that migrate such vast distances without landing that they begin to digest their own organs. There are glow worms that spend their entire life cycles in the dark and dank of caves.

Life has not merely occupied Earth; it isn't just a tenant. It has taken over the premises, transforming the planet's surface, atmosphere, and depths. Living things have put oxygen into the air and taken carbon dioxide out of the air, helping keep the climate of Earth in balance. And that doesn't even include us, the "mystery meat," the unknown quantity. We just got here and we have already set foot on the Moon and contemplated colonies on Mars.

In short, these three basic observations about life—that it is extremely robust across billions of years, that it is extremely diverse across millions of species, and that it is ubiquitous across the planet's many environments—are strong circumstantial evidence that life didn't merely show up: life seems to *belong* here.

Arrival of the Fittest

Let's take a closer look at what it might mean to say that life is written into the universe's building plan. In what ways? And why is it a useful idea?

Our understanding of life's diversity is anchored in Charles Darwin's theory of evolution through natural selection—his discovery that living things evolve incrementally, one into the other, over vast stretches of time. Nature acts on the variation that is naturally present in every population of living things, selecting those variants that produce the most offspring, or can better fend off competitors or predators, or are better adapted to their environment. What was once a nineteenth-century theory is now supported by such overwhelming evidence as to be part of our heritage for all time. Darwin's discovery of evolution through natural selection remains one of the most moving of all scientific breakthroughs—not least because it reveals that every living thing on Earth is our cousin. It unifies all life on Earth.

However, evolutionary biology remains an incomplete description of life, as Darwin himself knew when he published *On the Origin of Species* in 1859. That's because evolution does not attempt to explain how nature manages to conjure up such a variety of living creatures in the first place. When the Dutch botanist Hugo de Vries gave a series of lectures on Darwin's work at the University of California, Berkeley, in 1904, he concluded with a catchphrase: "Natural selection may explain the survival of the fittest, but it cannot explain the arrival of the fittest."[4]

Life-forms have to show up. Natural selection needs something to act on. Imagine that you are an employer looking for an employee who is able to do the work at hand and can fit in with the rest of your staff. You can select as carefully as you like, but if no applicants show up, your business is going to fold.

So where do nature's applicants for survival come from?

I've been fascinated by the extent to which leading scientists in the field have described this variation using words such as "chance," "accident," or "unpredictable," as though those words explained anything.

"Chance, pure chance" is how one of the pioneers of molecular biology, Jacques Monod, described the evolution of life. As a child growing up in the South of France, Monod became interested in biology when his father, an artist, read Darwin to him. Monod went on to a distinguished career at the Pasteur Institute and the Sorbonne. Among his many contributions, he elucidated the molecular mechanisms by which an organism's genes are turned on and off. He was awarded the Nobel Prize in Medicine and Physiology in 1965.

In his influential book *Chance and Necessity,* Monod laid out a manifesto based on the well-established observation that chance mutations in an organism's DNA help to drive changes in the organism from generation to generation. "It necessarily follows," he wrote, "that chance alone is at the source of every innovation, and of all creation in the biosphere. Pure chance, absolutely free but blind, at the very root of the stupendous edifice of evolution: this central concept of modern biology is no longer one among many other possible or even conceivable hypotheses. It is today the sole conceivable hypothesis, the only one that squares with observed and tested fact. And nothing warrants the supposition— or the hope—that on this score our position is ever likely to be revised."[5]

By "chance," Monod was referring to the chance mutations of genes, and to the other processes that shuffle genes within species and even across different species. But the word itself tells us nothing about why any of these shufflings produce something that is viable. The Swiss biologist Andreas Wagner has written engagingly about where life's innovations come from: "Where do the new variants come from that selection needs? One could answer this question with a vacuous platitude: New variants arise randomly, by chance. This platitude is still used today, but Darwin was already familiar with it. And he knew that it explains exactly nothing."[6]

Darwin wrote, "I have hitherto sometimes spoken as if the variations . . . had been due to chance. This, of course, is a wholly incorrect expression, but it serves to acknowledge plainly our ignorance of the cause of each particular variation."[7] When Darwin published *On the Origin of Species,* the molecular basis of life was not yet known; the very word *molecule* was not yet in use. But today, we can identify the molecular changes that give rise to evolutionary change.

There are two ways to probe the mystery of nature's innovation. One is to understand the pathways of evolution at a molecular level. If the machinery of life is set up to preserve itself faithfully from generation to generation, then how does it also manage to innovate? Does nature favor only certain pathways for evolution? Is that why similar structures and behaviors keep evolving over and over? (We'll explore this in Chapter 13.)

In addition, we need to understand why any of nature's innovations actually survive. Life has been so overwhelming successful, we have come to simply assume that evolution will come up with viable solutions to life's challenges. You can see this in the language that scientists have come to use in decribing evolution.

"Human and monkey brains have evolved dedicated systems for recognizing faces," wrote a reporter for the *New York Times,* "presumably because, as social animals, survival depends on identifying members of one's own social group and distinguishing them from strangers."[8] In other words, we evolved a new feature because we needed it. This, of course, glosses over the fact that nerve cells had to find a way to recognize faces, and they had to do it without any instruction.

A leading evolutionary biologist wrote that "it's no surprise that we evolved a completely erect posture to balance our big heads directly over our bodies—we didn't have any posterior appendage that could serve as a counterweight. If we hadn't become completely erect, we'd always be pitching forward, struggling to keep our balance."[9] Again, if we hadn't evolved this feature, we'd have been in trouble.

There is nothing wrong with this kind of writing; in fact, both authors are favorites of mine. But the reason that scientists know what this kind of shorthand really means is that nature is so profligate at coming up with solutions, so fecund in its diversity, that it seems nature can do virtually anything. So that has become the way that we talk about evolution. "X evolves Y, the better to do Z."

It is one thing to say that life is a chain of lucky events, but lucky comes in two flavors, including unlucky. I'm reminded of the cartoon showing a fellow on a blind date, proudly saying to the woman, "Well, I've been a failure in business, but I've had a string of successful marriages!" There are many more ways to fail in this world than to succeed!

And we have already seen that a molecule wants to be in the simplest, most stable state it can. The molecules of life are neither simple nor stable, and it takes a clever system to keep them around. The most-stable molecules belong to the world of eternal rest. That is one reason why that bottle of seltzer in Chapter 10 would still be here one hundred million years from now, while you and I and everything alive today will not.

Somehow, the trunk of the tree of life survives, even as the tree's leaves wither and die each generation. So any explanation of life needs to explain how life can maintain itself so far from equilibrium for so long. A successful understanding of life would have to show why *any life* survives.

The Time of Your Life

It is often said that the huge span of geologic time over which life evolved makes plausible the extraordinary complexity of life. "Given infinite time, or infinite opportunities," wrote the biologist Richard Dawkins, "anything is possible. . . . The large timespans characteristic of geology . . . turn topsy-turvy our everyday estimates of what is expected and what is miraculous."[10]

It is tempting to think of a living thing—a morning glory or a horse or a person—as somehow sculpted incrementally by the long passage of time. But, on closer examination, the vast span of geologic time does not really explain how complexity arises and why it is viable. After all, it takes just twenty minutes to assemble a bacterium from raw materials. A monarch butterfly takes a few weeks. A marigold takes forty-five days to mature. It takes only thirty-six weeks to assemble a human being.

It takes less time to produce a human being from raw materials than it does to assemble a sports car from scratch! You can argue that it took nature several billion years to evolve all the information needed to create a human being or an antelope. But the point is that the assembly and viability of any life-form does not rely on time. It relies on the astounding ability of molecules to assemble into the most-complex known structures with astonishing speed. The universe has the luxury of spending billions of years exploring the possibilities for life. But a living thing has to transact its business *now,* in the fleeting present.

Chemists distinguish between properties that depend on your state now and those that depend on the path you took to get there. For example, if you climb a mountain, your altitude at any moment depends only on where you happen to be standing. It doesn't depend on what route you took, and it has no memory of that route. But the total distance you traveled will certainly depend on the route you took.

In the same way, you are the function primarily of your genes, not how your genes got here. The story of life has to be understood in terms of the molecules that make life possible rather than accounted for with hand-waving about lucky accidents. And the molecular properties that make life possible are dictated by the infrastructure of the universe.

If you are skeptical, consider that we have already entered the era in which life-forms are divorced from evolution; we can bypass evolution entirely. Researchers have inserted human genes into mice, tobacco plants, and even a potato, where the genes work just fine. One can text the genetic information of a virus across the world and reconstruct it from scratch in the laboratory, modifying it at will. It is now possible to bring back animals that went extinct long ago, such as the passenger pigeon. It is possible to create new organisms that evolution could not achieve in the next million years. These developments underscore that life is ultimately sculpted by the properties of molecules, not time.

Today, of course, we understand well the molecular mechanisms that lead to diversity—nature's clever tricks by which genetic material is endlessly mutated, shuffled, and transferred from organism to organism, and even across species. Yet despite this understanding, the deepest and most disturbing questions remain unanswered: Who could have guessed that a motley crew of atoms could assemble themselves into the molecules of life? Into molecules that actually work? Into vast conglomerates that—unlike the molecules in a child's chemistry set—are able to cooperate at a level of such complexity that they can build nests, or raise their young, or ask questions? How did mere molecules manage to come up with an endless succession of solutions to life's challenges every day for three billion years? Why didn't life fail, when it had every opportunity to do so?

These may seem like rhetorical questions, but they are not. I suggest that the answers lie deeply embedded in the universe's building plan. Every property of every molecule is governed by the plan. Every interaction between molecules, and between groups of molecules, is governed by the plan. And while the environment provides surprises and contingencies and accidents galore, none of them have sufficed to derail the forward motion of life.

Eons from now, when our descendants have finally dotted the i's and crossed the t's of life, when the origin of life is well understood and the brain has been thoroughly mapped and its secrets at least partially revealed, I suspect that the wonder of life will remain undiminished. *How* life works will remain a source of intellectual excitement—but *that* life works will remain miraculous.

12

Information, Please!

A noted astronomer once suggested that the origin of life is so improbable that only by rolling the cosmic dice a nearly infinite number of times could life get under way. He noted that there are nearly an infinite number of ways to arrange the subunits of the molecules of life, such as RNA and DNA. How could nature ever find just the right sequence to form the first protolife?

He went on to say that the alternative scenario—that life arises inevitably from the universe's building plan, as John Archibald Wheeler conjectured—had a "whiff" of religion about it, because it would require that the chemical dice are loaded and that nature is somehow biased toward life.

How *does* nature find the right sequence of subunits for its DNA or proteins? How can nature innovate if it is constantly beset by chance and randomness? In this chapter, we will take a deeper look at the connection between chance and life. We will find that chance, far from being an enemy, is nature's way of creating the diversity on which all life depends. As strange as it may seem, chance and predictability turn out to be interwoven. They coexist.

The chemical dice are indeed loaded. We won't need to invoke an infinite number of universes to see how life might have started: one universe will do just fine.

Random Yet Predictable

Some years ago, while I was doing a project for Los Alamos National Laboratories, a researcher handed me a sphere of plutonium.[1] It was as warm as a puppy. And if I had held that sphere for a thousand years, rather than the ten seconds that seemed prudent, it would still be warm. The warmth comes from the energy released as plutonium atoms split apart into lighter fragments.[2]

This radioactive decay has a very curious property. If you were to observe an individual plutonium atom, you would have absolutely no way to predict when the atom would split apart. It might disintegrate tomorrow. It might split apart a thousand years from now. Or it might never decay! The moment of an unstable atom's demise can never be known—not even by nature itself. This is simply the way that nature has constructed radioactivity. Nature decrees this unpredictability. You might think that we could never predict anything useful about this system.

Yet when we look at the behavior of large numbers of plutonium atoms, we start to see a pattern. If only we had the patience to wait twenty-four thousand years, we would notice that half of the atoms will decay within that time. In principle, this half-life of plutonium can be predicted and calculated exactly, using the known laws of physics that are encapsulated in the universe's building plan.[3] We can also calculate the exact warmth that the sphere will produce, and a host of other, more sinister properties as well.

If you think about it, this is truly counterintuitive. The behavior of every single atom in the sphere is unpredictable, yet the behavior of the sphere as a whole can be accurately predicted. We'll see ways in which this counterintuitive idea holds true for the development of life as well.

This example gives us our first inkling that the concepts of randomness and unpredictability are more subtle than we may think. Somehow, chance and predictability can coexist. Let's see how this works in a living, evolving system.

Resist!

We can gain insight into how life innovates by observing evolution in action in a simple but striking laboratory experiment with bacteria. The researchers built a giant, rectangular petri dish, to which they added a colony of bacteria at each end. Across the length of the dish, they added increasing levels of an antibiotic known to kill the bacteria.[4]

As the bacteria grew and started to migrate across the dish, they encountered the antibiotic. Nearly all the bacteria died, but a very few mutants were able to grow slowly even in the presence of the antibiotic. (The mutants produced proteins that happened to counteract the antibiotic.) As these resistant strains grew and migrated, they encountered a region with ten times the lethal dose. This slowed their growth, but again a few mutants survived even this dosage and went on to reproduce. Within days, descendants of the original bacteria were able to grow in one hundred times the normally lethal dose of the antibiotic.

This experiment is a beautiful illustration of evolution by natural selection: you start with random mutations, you subject the organism to a demanding new environment, and you wind up with survivors that bear a new trait—in this case, antibiotic resistance. At first glance, the experiment seems to bear out Jacques Monod's dictum that "chance, pure chance" is at the root of evolutionary innovation, since the bacteria's random mutations ultimately led to the new trait.

But on closer inspection, the experiment is evidence for the opposite conclusion as well! The bacteria from *both* sides of the dish make it to the center; both groups are able to come up with mutants resistant to the antibiotic. In fact, in this particular experiment, at least six different mutant colonies were able to make it to the center of the dish, which contained the highest antibiotic levels. Furthermore, if the researchers repeated the experiment three, four, or twenty times, using the same starting bacteria, the outcome would be the same: an antibiotic-resistant strain would evolve. The specific mutations might be different, and the time it took might be a tad shorter or longer, but the outcome would be the same trait. So, for this particular experiment, the evolution of antibiotic resistance is not an accident; in fact, it is predictable and repeatable!

Remarkably, this process is analogous to the plutonium sphere's radioactive decay: it is unpredictable at the molecular level, and yet the behavior of the overall system is highly predictable.[5]

Who gets credit for the bacterium's discovery of antibiotic resistance? Chance, or the bacteria's protein molecules? The answer is both. Chance is necessary to ensure a *diversity* of proteins; if the bacterial population has enough diversity, then one of its members will survive the effects of the antibiotic. But the lion's share of credit goes to the protein molecules that are able to neutralize the antibiotic. And the properties of those heroic proteins are dictated not by chance but by the universe's building plan.

There is an important take-home message from this experiment, and it lies at the heart of the story of life: nature doesn't have to invent precisely the right molecule to accomplish a particular task. Many different proteins can accomplish the same task—in this case, conferring antibiotic resistance. In this particular experiment, nature did not need to provide any additional information to the bacteria.

Chance is the instrument for generating a diversity of molecules. But the music resides in the properties of the molecules.

Intuition Is Not a Reliable Guide to What Is "Random"

The mathematician Gregory Chaitin created a puzzle that illustrates an unusual property of chance and randomness. We won't have to solve his puzzle, but it is worthwhile to understand the problem and its unexpected result.

Draw a triangle with equal sides inside a circle, as shown in Figure 12.1. Now draw a couple of chords inside the circle. (A chord is a straight line that cuts across the circle.)

Here's the puzzle: What is the likelihood that a chord drawn at random will be longer than the side of the triangle? That is, if you draw lots of chords at random inside the circle, what fraction of them will be longer than the side of the triangle? For example, in Figure 12.1, the line A is shorter than the triangle side, but the line B is longer.

In trying to solve what seems like a math puzzle, Chaitin first had to figure out how one would draw a chord at random. It sounds simple, but what does it actually mean? He described several ways to do it. You could attach a spinner from a board game to the edge of the circle, and the spinner would sweep out chords of every length. Or you could roll a broomstick across the circle to create

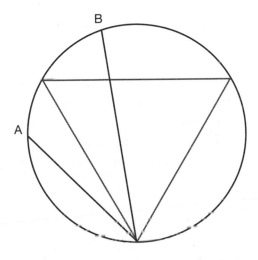

Figure 12.1. A triangle with equal sides is drawn in a circle. What is the likelihood that a chord drawn at random (such as A or B) will be larger than a side of the triangle? The answer may surprise you.

a chord that way. Or you could drop a pencil onto the circle and create a chord that way. There are many ways to randomly draw a chord.

To Chaitin's amazement, he found that each one of the methods gives a different answer! (The answer for the three methods described turns out to be one-third, one-half, and one-quarter, respectively.) There is not a unique answer. It all depends on how you propose to draw a chord "at random."[6]

What seemed like an innocent question has a less than innocent answer. Our intuition about randomness is not a good guide to how the world actually works. Could it be that chance and randomness are not the most useful concepts in analyzing the biological world as well?

"Information" Is the Concept We Need

In our quest to understand how life might be written into the laws of nature, we will find that a modern concept, information, may be more useful than the concepts of chance and randomness.

To see how this works, let's start with an intriguing puzzle, also based on the work of Chaitin. Examine the two following numbers. Which of them would you say is more likely to occur in the real world?

A. 2.3254779852211
B. 2.2222222222222

Take a moment to think about it. If you just happened to come upon a number during the course of your day, which one do you think you would be more likely to encounter?

Many people tend to say that A seems more likely. The sequence of digits looks like something that might actually come up in a lottery, whereas number B seems out of the ordinary. How likely could it be for a number to have all twos?

A math teacher might have a different answer. Both numbers might be considered equally likely. If you chose each digit at random—for example, by picking a digit from 0 to 9 out of a hat—then each number would have the same chance of being drawn. (Yes, it is unlikely to pick all twos. But it is *equally* unlikely to pick the specific sequence of digits in A.) For both A and B, you would have to guess right for every digit if you wanted to get that specific sequence. The math teacher might say that we are misled by our assumption that a random-looking pattern is more common than one that looks special.

Surprise! There is an entirely different way to look at this question, one that corresponds more closely to the real world. From this modern perspective, B is the more likely number to occur in the real world.

Here's why. Instead of invoking chance and randomness, you could ask, How much information must I provide in order to create each of these two numbers? For number A, you have to specify every digit. If the number were a hundred digits long, you would have to specify all one hundred digits. How else would anyone know what they were? If the number had an infinite number of digits, you'd have a lot of specifying to do! So number A requires a great deal of information to create or reproduce. But for number B, you would only have to say, "Two, followed by twos all the way out." It takes much less information to create number B.

It turns out that the less information you need to specify something, the easier it is to create—and the more likely it is to occur in the real world.

This makes sense. We have all had the experience of asking a stranger for directions. If the answer involves turning left, then right, then right again, and then left and looking out for the supermarket, the gas station, and the old house with tulips in front, the chances are that we will wind up just searching at random; who could remember all that information? But if the stranger tells us, "Second light, take a left, you'll see it," then chances are you will actually use that information.

You can test this idea easily. Just Google the two numbers to see which one occurs more frequently in the real world. You'll find that number A (2.3254779852211) doesn't come up at all.

That's because it requires so much information to specify than no one besides me has come up with it; the chances of its cropping up in the real world are simply too small. (If it does come up on an Internet search, it will only be in connection with this book.)

But number B, with all the digits 2, comes up thousands of times in a Google search. It takes less information to create this number, since you don't have to specify each digit. You have a simple process for writing the digits down. In the real, physical world, that makes number B more likely to come into existence.

We're onto something here. Forget about how random something appears and ask instead how much information you need to create it.

Chaitin considered a more subtle puzzle. Which number is more likely to appear in the real world, A or B?

A. 2.3254779852211
B. 3.1415926535897

Both of these numbers have digits that form a completely random sequence, so you might think that they both require the same amount of information to create. But notice that B is not just any random number. It is the number pi, defined as the distance around a circle divided by the diameter of the circle.

In order to create B, you don't need to specify all of its digits: you need only specify the *process* that produces its digits. For example, you could say, "Draw a circle and divide its circumference by its diameter. The result is the number pi." That's a lot shorter than having to specify every digit of pi out to infinity! The process may be difficult to carry out—how do you draw a perfect circle?—but it takes just a few words to describe. (If you don't want to be bothered drawing circles, there are other simple procedures that generate as many digits of the number pi as you would like, without specifying in advance what they are.)

Number B shows up on a Google search thousands of times. It is also likely to arise everywhere in the universe that a technological civilization has emerged and uses circles; eventually they will discover the number pi. The universe may be swarming with it!

This little example is analogous to the plutonium sphere. Each individual digit appears to have been chosen randomly; there is no pattern to the sequence of digits. But the number as a whole is not random; there is a simple and natural process that generates it.

Rephrasing Wheeler's Question

We can now make more sense of John Archibald Wheeler's question by rephrasing it using the idea of information. Instead of asking whether the universe's building plan "guarantees" life, we can ask,

> Does the emergence of life in the universe require any additional information, beyond that contained in the universe's building plan?

We have already seen that no additional information is required to produce the physical universe. In that sense, the universe is like a windup toy: just start the universe, and it will automatically create a glittering array of galaxies, stars, and planets scattered across the night sky. Guaranteed. No tools needed!

The more specific you choose to be, however, the more information you require. For example, will it be raining exactly two hundred years from today? Yes, the universe's building plan guarantees that it will be raining then—but that rain may be on a planet on the other side of the galaxy. If you wanted to know whether it will be raining in your particular hometown, right here on Earth, you would need additional information beyond that contained in the building plan.

We can apply these ideas to an example from our own bodies.

The Cleverness of Molecules

The human immune system is fiendishly complex, so we'll focus on just a portion of it, and even then we'll need to simplify further. This example will highlight why it makes more sense to analyze life's processes in terms of information, rather than chance.

At any given time, your body circulates roughly a trillion different immune cells. Each cell produces a unique protein called an antibody. The immune system's strategy is to produce so many different antibodies that at least one of them will have a shape that can attach to a foreign invader such as a virus or bacterium.[7]

But there is a challenge: How can the body produce a trillion different proteins? Normally, a protein is specified by a single gene; one gene produces one kind of protein. Your body contains thousands of genes, but nowhere near a trillion genes. So how can the body create this incredible diversity of antibody proteins?

The secret is that the body doesn't have to provide the specific information for producing each antibody. Instead, it has mechanisms for generating enormous diversity. Some of these mechanisms involve chance. Here's a glimpse of how it works.

Unlike most proteins, an antibody protein is derived not from a single gene but rather from many small segments of different genes. These snippets of genes are then stitched together to form the gene that will produce the antibody protein. Just as there are many ways you could string beads together to form a huge number of patterns, so too are there a huge number of ways to put the few snippets together. So, right away, the body generates much more diversity than it starts with.

Normally, cells have very precise mechanisms for recombining genes. But the immune cells arrange for the process to be sloppy and error prone; letters are either added to or deleted from the ends of the gene segments, leading to whole new "words." That sloppiness also creates diversity.

The immune cells whose antibodies happen to bind to the invader are stimulated to mutate, in the hopes of fine-tuning the antibody and finding a similar one that binds the intruder even more strongly. This hypermutation is a million times faster than the normal rate at which a gene mutates. Why settle for business as usual when you are under attack by invaders?

The lucky immune cell with a "winning" antibody then devotes itself to spewing out copies of the antibody at the enormous rate of two thousand per second. This puts such a strain on the cell's protein-making machinery that it can no longer divide or grow.

To accomplish all this, the immune system is exquisitely controlled by other proteins and genes. One system ensures that an antibody is produced only after the gene rearrangements have taken place. Another system detects whether an immune cell has an antibody that attacks *you* instead of the foreign invaders. If so, the cell is given a chance to mutate further; otherwise it is destroyed. Still another system enables some antibodies to cross the placenta in order to protect the mother's fetus.

Jacques Monod saw the immune system as the apotheosis of chance. "It is indeed remarkable to find chance at the basis of one of the most exquisitely precise adaptation phenomena we know. But . . . only such a source as chance could be rich enough to supply the organism with means to repel attack from any quarter."[8]

He went on to write that the immune system appears "miraculous to some" due to "the extreme difficulty one has in imagining the inexhaustible resources of that ocean of chance upon which selection draws." He continued, "When one ponders on the tremendous journey of evolution over the past three billion years or so . . . one may well find oneself beginning to doubt again whether all this could conceivably be the product of an enormous lottery presided over by natural selection, blindly picking the rare winners from among numbers drawn at random. [Nevertheless,] a detailed review of the accumulated modern evidence [shows] that this conception alone is compatible with the facts."[9]

But chance and luck and lotteries have little to do with the marvel that is the immune system. The immune system is about diversity, not chance. An extremely sophisticated system of genes harnesses chance to create that diversity.

The immune system illustrates one way that nature can produce enormous diversity from a relatively small amount of information: just a few genes can create a trillion different antibodies. Your body doesn't have to specify the details of those antibodies ahead of time, such as their amino acid sequence. It can cede the task to chance, because there is no magic formula for the right antibodies; any trillion antibody proteins will apparently contain some that will do the trick. That is why you and I can both have immunity to the flu, yet your antibodies will not be identical to mine.

Nature is a virtuoso, a true impresario, whose talent lies in the fantastic properties of molecules. We can see how the immune system works, we understand it, we control it, yet it still remains remarkable that mere molecules can create such a system.

The Mystery of Origins

Where did the immune system come from in the first place? If we trace life all the way back to the origin of life's infrastructure, the mystery only deepens. We hit a wall of ignorance.

We have no direct evidence of what happened more than three billion years ago. And life's infrastructure itself poses a paradox: DNA contains the information needed to create proteins, yet DNA also needs proteins in order to function in the first place. Which came first? How did it work?

The origin of life remains perhaps the deepest problem in science—right alongside the origin of the universe. You can tell how difficult the problem is from the way scientists have discussed it over the years. Francis Crick, the codiscoverer of DNA's structure, suggested in the early 1980s that "the origin of life appears at the moment to be almost a miracle, so many are the conditions which would have had to have been satisfied to get it going."[10] The physicist Philip Morrison once said that he looked forward to the discovery of even the simplest life-form beyond Earth, because that would at least transform the origin of life from a miracle to a statistic.[11]

The problem is so difficult that many reknowned scientists considered the possibility that life *didn't* originate on Earth—that it arose elsewhere and made its way here, perhaps as microbes stowing away on comets and asteroids. This speculation, called panspermia, was at various times embraced by Crick and by the astronomer

Fred Hoyle. The Nobel chemist Svante Arrhenius (who you'll re-call presented Albert Einstein with the Nobel Prize and who was one of the first to describe global warming) was also a proponent of the theory. But as early as 1872, the British chemist Robert Angus Smith dismissed the panspermia theory on the grounds that it merely swept the problem farther from Earth, the one place that seems most hospitable to life.[12]

Today, the study of the origin of life is flourishing in laboratories around the world. The field has been revived by two spectacular breakthroughs: One is the revolution in molecular biology, which is providing tools for probing and even designing life at the mo-lecular level. The other is the discovery of thousands of solar sys-tems far beyond our own, raising the possibility that the discovery of life in our galaxy is just around the corner (see Chapter 16).

Let's look at two examples of research that bear on the con-nection between chance and the likelihood of life.

Evolving Proteins in a Test Tube

What would happen if you could assemble a trillion different protein molecules, chosen completely at random, with no connection to any living thing? Would any of these proteins have properties that were useful to life?

You might think the chances would be extremely unlikely. After all, there are nearly an infinite number of possible proteins that a living cell could produce.[13] There are so many possibilities that you could fill the observable universe with one example of each of them. No life-form could ever have manufactured even an infinitesimal fraction of the possibilities. So, how did life manage to find the proteins that are useful and avoid the ones that aren't?

This was the question that Jack Szostak and his colleagues investigated in a beautiful series of experiments at Harvard Medical School. First, they generated one trillion different protein molecules, produced randomly with no connection to any living thing. That may seem like a huge number, but keep in mind that molecules are so small that you can't even see a trillion of them with the naked eye. And one trillion is merely an infinitesimal fraction of the number of possible protein molecules. The important point is that, because the molecules were constructed at random rather than from a living cell, they were a sample of the vast repertory of all possible proteins, not just the ones we see now in nature.

The researchers then asked the question, Are any of these random proteins able to carry out a simple function needed by a living cell? The function they chose was the ability to bind the high-energy molecule that powers the basic chemical reactions in every living cell. The molecule is called ATP (short for *adenosine triphosphate*); the name sounds esoteric, but ATP is as fundamental to life as gasoline or a battery is to cars on the road. Proteins that bind ATP include the ones that carry out photosynthesis, that contract your muscles, and that help cells divide. Binding is important: if you want to use a hammer to build something, first you have to be able to hold the hammer. A bad grip will not serve you well.

Amazingly, Szostak and his colleagues found that, among his random sample of proteins, there were indeed a handful that bind ATP. Some of these molecules even hold on to ATP just as effectively as the proteins found in nature.

The researchers then investigated whether these proteins could evolve further. Could they acquire a second useful function without losing the ATP-binding function? For the second function, they looked for whether the protein could maintain its three-dimensional shape at high temperature. (This function is important for micro-organisms that live in hot springs or their sea-floor equivalents. The latter may have played an important role in the origin of life.) At high temperatures, proteins are buffeted more strongly and can lose their normal, three-dimensional structure.

Remarkably, the researchers found that they could indeed select for proteins that had both functions. We conclude that useful proteins are plentiful, not rare: you can find them even in a random assortment of proteins. To put it another way, many different proteins can accomplish the same useful task; that must be the case, or we would not have found a useful protein in such an infinitesimally small sample.

These experiments provide further evidence that life can arise naturally from the universe's initial conditions. If nature came with a manual for living things, one of its instructions would be simple: keep making a diversity of proteins. The proteins will do the rest. Life doesn't need a magic formula for the amino acid sequence of a useful protein. It doesn't need an infinity of universes to keep rolling the cosmic dice. The dice in our own universe are already loaded.

Of course, these experiments only tell us about how one or two biological functions can arise. A couple of useful proteins do not constitute life. And since proteins cannot reproduce themselves, they would not be able to sustain life. To understand how mere molecules might reproduce themselves, we need to turn our attention to the world of RNA.

Entering the RNA World

To probe some of the steps that might have led to the origin of life, Gerald F. Joyce and his colleagues at the Scripps Oceanographic Institute have actually been evolving RNA molecules in a test tube. Joyce is one of the leaders in this field, and he has received numerous awards for his work.

When he was a student at the University of Chicago, Joyce was influenced by the novels of Thomas Pynchon, especially *Gravity's Rainbow*. As Joyce described it, the book depicted a century "dominated by physics, and of human behavior swept along by the inescapable tide of physical laws," especially the rush to war and disorder. "It is a distressing picture," Joyce wrote, "made more chilling by Pynchon's observation that rather than resist the universal tendency towards disorder, humans have become highly adept at promoting it."[14]

What struck Joyce was Pynchon's idea of a "counterforce"—some kind of organizing principle that opposes the universe's long slide toward chaos. Joyce found his own counterforce in Darwin's idea of evolution based on natural selection—appropriate for a force that Pynchon had called "the green uprising."

In choosing a career path, the young Joyce decided that his "task was clear: starting from simple chemical building blocks, construct a self-reproducing system that, like DNA-based life on Earth, would be capable of undergoing Darwinian evolution."[15] Of course, no one has recreated the origin of life in a test tube yet, nor is anyone likely to in the near future. But Joyce and his colleagues have taken the first promising steps. They are exploring the idea that the first living things might have been based on RNA.

This would have happened long before DNA and proteins became the foundation of life's machinery.

An ancient RNA world would make sense, because RNA combines some of the properties of both DNA and proteins. Like DNA, it can carry genetic information. And like a protein, it can also catalyze chemical reactions.[16] So RNA might be a bridge between the DNA and protein worlds, and it might have served to jumpstart life.[17]

Joyce and his colleagues have been able to engineer RNA molecules that reproduce themselves. In fact, the molecules can evolve in the laboratory to become more efficient at reproducing themselves.[18] This is Darwinian evolution in a test tube: if you are a molecule that can outcompete other versions of you by producing more copies of yourself, then you can dominate the test-tube landscape. The experimenters need only ensure that there is a diversity of RNA molecules in the test tube and that new raw materials are replenished periodically. The molecules themselves do the rest.

The researchers went on to develop an RNA whose ability to reproduce depends on its binding a small molecule. For their experiment, they used a caffeine-like substance found in cocoa beans and brewed tea, called theophylline.[19] These experiments support the plausibility of an RNA world filled with molecules that could perform useful functions and also reproduce themselves.

Joyce is modest about his achievements. "The counterforce, it turns out, is not much to look at," he notes, "typically 20 microliters of a clear, colorless solution. But in those solutions the Titans are rumbling. We can begin in the lab on a Monday with a population of 100 trillion random-sequence [RNA] molecules and by Friday witness the emergence of order in the form of [RNA molecules] of a particular sequence that perform a specific catalytic task." The result, he says, "has created an island of order in a universe that is forever tumbling toward a state of maximum entropy."[20]

Now we can see the connection between chance and life more clearly. Chance is the engine of diversity, and with enough diversity anything seems to be possible. Nature says, Give me a trillion bacteria, and I'll guarantee that one of them has resistance to your antibiotic. Give me a trillion antibodies, and I'll guarantee that one of them fights the invader. Give me a trillion protein molecules, and you can count on finding the function you are looking for. Give me a trillion RNA molecules, and we are off to the races.

In these examples, chance is the handmaiden of predictability and reliability.

Chance and Life

The closer you look at nature, the more you will discover the myriad ways in which life harnesses chance to ensure its survival. A gypsy moth flits randomly this way and that to avoid being snatched out of the air. The seeds of a plant germinate at chance intervals to better ensure that the seedling will emerge when conditions happen to be favorable.

I want to end this chapter by revisiting the same phenomenon with which we started: radioactivity. Could life on Earth possibly depend on something as relentlessly random as radioactive decay?

Remarkably, it appears so. Consider the decay of uranium atoms deep within the Earth. The heat from this radioactivity accumulates over eons, eventually melting iron and rock. It provides the energy that carries great continents on their slow drift across the planet. And it helps to drive the geologic cycles on which the

Earth's climate depends. Though it is by no means certain, the process may be essential for life to prosper.[21]

Were we to look at the radioactive decay of just a few uranium atoms, we would see a process that was the very epitome of chance. Yet the big-picture view shows that the countless uranium atoms inside the Earth provide an entirely reliable, continuous source of energy. The half-life of its most common unstable form is well matched to the long times required for the evolution of life. The half-life, abundance, and radioactive decay of uranium are all right there in the universe's building plan. Guaranteed. No additional information required!

13

Is Evolution Predictable?

At the start of a new century, in the year 1800, the explorer and naturalist Alexander von Humboldt left the coast of Venezuela and ventured inland on a 1,700-mile journey of discovery up the Orinoco River. Humboldt and his companion, the botanist and physician Aimé Bonpland, would spend four months in this wild and dangerous region. The team would eventually discover the confluence of the Orinoco and Amazon Rivers; test the powers of the bark of the cinchona tree, from which quinine is derived; and make contact with several indigenous groups.

But the part of Humboldt's adventure that intersects our story of the universe is his terrifying encounter with one of nature's oddities: the electric eel, whose discharge can kill a grown man. The Indians took Humboldt and Bonpland "to a large reservoir of slimy water, surrounded by odorous plants," according to an account published in 1852.[1] The Indians could not use fishing nets to capture a specimen for Humboldt, since the eels buried themselves in the slime, so they used another technique.

They captured wild horses and mules and drove them into the water. The stamping and commotion drove the eels to the surface. "They swim on the surface of the water," the account continues, "and press themselves on the belly of the mules and horses."

> A strange combat now begins; the Indians, provided with long thin bamboo canes, encircle the ditch. . . . By wild screams and threats with their long canes, they prevent the horses coming ashore and escaping. The eels, terrified by the noise, defend themselves by the repeated discharge of their electric forces.
>
> In the first five minutes two horses were already drowned. The eel, which is five feet long, presses against the belly of the horses, and discharges its electricity along its whole length, which stuns the abdomen, entrails, and heart of the horse. The horse sinks down exhausted, and is drowned, because the continuing struggle of the eels with the other horses prevents its rising again.

Humboldt feared that all the horses might die, but "the violence of the unequal struggle gradually abated, and the exhausted electric eels dispersed; for they require long rest and abundant nourishment to regain the strength spent by the frequent discharge of their electric organs. The horses and mules recovered from their terror, their manes no longer bristled, and their eyes no longer glared so fearfully." By the next day, Humboldt was able to capture a specimen of the exhausted eels.

When Humboldt visited the United States a few years later and met with the president, Thomas Jefferson, he avoided mentioning his harrowing episode with the electric eels, choosing instead to describe the teeth of a woolly mammoth that he had discovered near the equator. But his strange encounter with the eels stayed in his mind and would appear in his writing on several later occasions.

The year that Humboldt died, his friend Charles Darwin published *On the Origin of Species,* which deepened the mystery of the electric eel. "The electric organs offer another and even more serious difficulty," Darwin wrote.[2]

You might think that a creature as odd as the electric eel would be a one-off, a one-time invention that nature might be proud of and then move on. But Darwin was mystified that there were at least a dozen different species of electric fish, and strangest of all, none of them appeared closely related to each other. He noted that there were many cases in which plants or animals had evolved similar structures. Similar, but not identical. This led him to believe that the structures *must have evolved independently of each other:* "I am inclined to believe that in nearly the same way as two men have sometimes independently hit on the very same invention, so natural selection, working for the good of each being and taking advantage of analogous variations, has sometimes modified in very nearly the same manner two parts in two organic beings, which owe but little of their structure in common to inheritance from the same ancestor."[3]

But how in the world is this possible? If a variation arises purely by chance, then why should it appear again and again?

Fast-forward a century and a half to our own time. Biology's marvelous molecular tools have enabled us to scrutinize the structure and development of living things at the level of their DNA— and to use the DNA to draw inferences about how the organisms evolved.

Researchers have found that the electric organs in fish have been reinvented at least six times.[4] These organs evolved from what were originally muscles. The makeover was not a simple kitchen renovation; it required wholesale changes in the development of the fish. Nature had to find ways to shut down the contraction of the original muscles; to make the cells larger; to stack the cells in an array so as to increase the voltage, like stacking the batteries in a flashlight; and to make certain the electricity would be conducted into the water and not merely dissipated inside the fish. It turns out that even though the organs of the different electric fish species look different, they all arise from remarkably similar developmental pathways as the fish matures.

Convergence

A lively debate has swept across evolutionary biology. To what extent is evolution unpredictable and governed by chance? And to what extent are the results of evolution predictable and predetermined by nature?

If life is truly on the universe's to-do list, then we might expect similar forms of life to emerge repeatedly. We can't test this idea by replaying evolution from its very beginning, but nature has been carrying out experiments that are almost as revealing. We can look at the immense tapestry of life and ask, Has evolution often converged on similar forms of life? The answer is yes. In fact, there are now hundreds if not thousands of examples of convergent evolution.

In this chapter, we'll look at just a few of life's innovations that have appeared repeatedly, like a recurring motif in a song. The cumulative impression we will get is that life is like a river that can run this way and that but which eventually carves out channels that seem to guide its course. Certain of life's structures, functions, and behaviors seem to bubble up insistently. So insistently that biologists have already begun to pose the question, Is evolution predictable?

Thanks to the ongoing revolution in biotechnology, it is now possible to explore some remarkable examples of convergent evolution at the molecular level, and begin to understand the factors that may drive it.

A Light in the Dark

Convergent evolution has many causes, some of which are not relevant to our story. Sometimes convergent evolution merely reflects the environment's constant, merciless selection of the fittest—not some innate tendency for certain traits to evolve. This is often the case when a particular trait already has a range of values that selection can act on. For example, indigenous people in tropical Africa, Asia, and South America tend to be shorter, presumably because in these regions, shorter means fitter. People living at higher altitudes tend to have hemoglobin that binds oxygen more strongly. Again, it's not that these traits evolve repeatedly. They preexist in the population, then spread more widely in environments where they are favored.[5]

But when nature does repeat itself, the results are extraordinary. We learn in school that the eye has been reinvented many times across the animal kingdom. But it has even been reinvented *in reverse:* Some squid not only have eyes that capture light, they also have organs that produce light and shine it back out into the ocean. This involves creating crypts along the bottom portion of the squid's body, then filling the crypts with a species of luminescent bacterium that inhabits the ocean. The squid has developed elaborate mechanisms for feeding the bacteria, and for ensuring that only the correct species of bacterium is nurtured. The light-producing organs use some of the proteins normally found in the eye, even to the point of creating lenses that concentrate the light in the desired direction. The squid goes to all this effort in order to camouflage itself; from below, the light serves to hide the silhouette of the squid against the lighter sky above the ocean.

Remarkably, this arrangement has evolved multiple times in squid that are only distantly related to each other. Even though there are differences in the details of these organs, they are found to have arisen through similar changes in the genes.[6] Among fish, bioluminescent organs have arisen independently at least seventeen times.[7]

Now Hear This

An equally remarkable example is the convergent evolution of the middle ear in mammals.[8] We mammals are distinguished by having three bones in our middle ear, whereas reptiles and birds have only one bone there. Where did the two extra bones come from? It turns out that as some reptiles evolved into mammals, two of the bones from the reptile jaw wound up migrating to become part of the mammalian middle ear. This evolutionary adaptation helped mammals to hear better than their reptilian ancestors.

You might think such a fortuitous change would occur once and never again. But evidence indicates that it actually happened at least twice. The strange group of mammals that includes the platypus evolved from a reptilian ancestor separate from that of most other mammals, yet it gained the two extra bones that wound up in the inner ear. In this mini-replay of evolution, the outcome was the same.

Even more remarkable is that some insects evolved ears that arrived at the same engineering solution as mammals' ears.[9] This was accomplished with entirely different materials, of course, since insects have no bones. Here is the problem and its solution: When animals first emerged from the sea and colonized the land, they had a rude awakening: sound is transmitted much less efficiently through air than it is through water. If you wanted to stay alive, you had better develop a more sensitive ear than was needed underwater. Fortunately, nature came up with a marvelous engineering solution.

In mammals, this clever ear consists of three parts. The outer part is the eardrum, which vibrates as it captures sound waves. The vibrations are transmitted to the middle part of the ear, where those three little bones serve as levers that amplify the vibrations and transmit their force to the inner ear. This inner sanctum contains an array of cells, arranged almost like the keys of a piano, that analyze the sound according to its frequency. Cells at one end respond to lower frequencies, and cells at the other end respond to higher frequencies.

Amazingly, a tropical katydid boasts an arrangement that is virtually identical in design to the mammalian ear. (Katydids are those brilliant-green insects that resemble grasshoppers.) The katydid's "ear" resides in a joint of one of its legs. But, like the mammalian ear, it has three parts: an outer eardrum, a middle ear with a bone-like structure that transmits and amplifies the vibrations, and a fluid-filled inner part that analyzes the frequencies. There is even an array of sensing cells laid out like the keys of a piano, just as in the mammalian ear.

Since katydids and humans are separated by hundreds of millions of years of evolution, one can only marvel that this replay of evolution produces such similar structures—even though the construction materials and the location of the ear are completely different.

Ants and Hyenas

If you examine enough different species, you start to see remarkable similarities that you wouldn't have thought were there. I happened to notice online that the mandible (jaw) of a fire ant is remarkably similar to the jawbone of a hyena. Again, the materials that nature has to work with are completely different in the two cases; the ant jaw is made not of bone but of chitin, a polysaccharide related to cellulose.

The explanation for coincidences like these ultimately lies in the genes and molecules that determine an organism's body plan. It turns out that a relatively small number of genes can produce a huge number of variations on the structure and arrangement of body parts. In part for this reason, living things share a much larger proportion of their genes than one might have imagined. More than half the genes found in your body, for example, are also found in fruit flies. They are simply deployed differently.

You are not in danger of sprouting wings, however. Nature recycles the proteins it has come up with and repurposes them for different uses. For example, there is a protein in fruit flies that helps to determine the placement of the fly's antennae. Mutations in this protein can lead to a fly that has a leg on its head, or that has an antenna where its leg should be. Yet virtually the identical protein is found in you and me, where it helps to determine where our bones should develop. If you inject the fly protein into a mouse, a bone will begin to form at the injection site.

A group of these proteins, called bone morphogenetic proteins (BMPs), has played an important role in evolution and also in the study of evolution. Take Darwin's finches—fourteen species of songbirds found in the Galapagos Islands. During his 1835 expedition to the Galapagos, Darwin observed that the beaks on these birds varied greatly from species to species; moreover, each species had a beak that was well adapted to gathering the particular food on its island. Darwin realized that, on each island, nature had favored the species whose beak shape was best able to exploit the food on the island.

As we've seen, survival of the fittest does not explain the *arrival* of the fittest. But thanks to laboratory experiments, we can now glimpse that arrival. We now know that the beak shape of a finch is determined largely by the timing and amount of a single protein, called BMP-4.[10] You can even tweak the beak shape of a developing chicken by controlling this protein.

In short, even a limited system of molecules can give rise to a huge variety of anatomical structures in insects, birds, mammals, and other creatures. A parsimonious universe still in creation!

The Cactus That Reversed Course

One of my favorite examples of convergent evolution is a cactus pretending to be another succulent. The cactus is *Leuchtenbergia principis* (Figure 13.1), which you might mistake for an agave or aloe plant. The story of this impostor is very strange.

Most plants have leaves that are flat and plentiful, in order to collect as much sunlight as they can. But for desert plants, sunlight

Figure 13.1. The cactus *Leuchtenbergia principis*. What look like leaves are actually part of the stem, which evolution has fashioned into the same shape as the leaves of aloes and agaves. Credit: Photograph by the author.

is not the issue; water is. Succulent plants are a response to this challenge. An agave, for example, has just a few, thick, tough leaves that store water. Cactus plants are an even more extreme adaptation; they have lost their leaves entirely. In fact, the spines on a cactus are the only remnants of what used to be leaves. Over eons, the plant's chlorophyll migrated to the stem; the body of a cactus is actually its stem. The really successful cactuses became globular or barrel-shaped, because a sphere is the shape that has the least surface area for a given volume, and therefore the least opportunity for water to escape from its pores.

But what happens when the environment changes again and a cactus would rather have its leaves back? With *Leuchtenbergia,* nature has fashioned what look like leaves from the stem of the plant. It is a remarkable deception. It is as though you had once lost your arms and legs and spent many generations living life as a torso, but then nature fashioned new arms and legs, fingers and all, from your liver.

Adding to the marvel is that cactuses retain their memory of the ancient days when they used to be normal plants. When I was a child, in the days before the measles vaccine, I contracted the measles and missed the biology class in which the teacher discussed plants that have two seed leaves, called dicotyledons. For years afterward, I would wonder, What in the world are seed leaves, and why would anyone care about them? I felt deprived.

But when you grow a cactus from a seed, you see what the fuss is about. When the seed germinates, two normal little leaves appear; you think that you may have planted a tomato by accident. But then the growing stem reveals itself as the cactus. The two seed leaves fall off, having nourished the plant and gotten it off to a good start. Or maybe the leaves are merely there as a last vestige of the cactus's parentage. Either way, the leaves are a marvelous reminder that we are all part of a continuous, ages-old chemical reaction that connects all living things and has never stopped weaving its magic.

Swimming and Flying Underwater

It is striking enough when the same shapes emerge repeatedly. But even more remarkable is the convergent evolution of animal behaviors. A good example is the swimming behavior that has evolved independently in tuna and some species of shark, such as the mako shark. These species are not buoyant, so they have to swim constantly for their entire lives in order to stay afloat. They have an unusual physiology that catapults them into the ranks of the best swimmers. Unlike most fish, for example, they maintain a body temperature well above the surrounding seawater, and they have evolved a clever heat-exchange system that preserves that body heat.

Most importantly, they have a highly efficient technique for swimming. Most fish undulate from the midpoint of their bodies; the waves of body motion propel a fish forward but also increase the drag force as the fish moves, wasting energy. By contrast, a tuna keeps most of its body still, directing the force of its muscles to its tail. As a result, it maintains a sleek profile and can speed through the water at more than thirty miles per hour.

Researchers note that "the evolutionary convergence between [these species] is so striking that in many ways these distantly related groups resemble each other more than they resemble their own close relatives."[11] This is all the more remarkable given that the two lines of fish have had no common ancestor for more than four hundred million years.

Another example is the tiny sea snail *Limacina helicina,* also called the sea butterfly. Like many other forms of zooplankton (animal drifters), the sea butterfly swims to the ocean's surface every night to avoid predators and to seek its own food. As strange as it might seem, this creature uses its two appendages to literally fly underwater, using the same complex motions that small insects use with their wings. These motions not only propel the snail forward; they generate lift and keep the snail "airborne" under water.[12]

The sea butterfly is weighed down by its dense shell, so it needs a more powerful way to lift itself through the water than the simple paddling motion that marine organisms typically use. To use its appendages as wings rather than oars, the creature must flap them in a complex figure-eight motion and add an odd twist of its body, allowing it to soar underwater as though it were a fruit fly maneuvering through the air. No one yet knows the genetic mechanisms that enable a sea butterfly to fly underwater. But given that sea butterflies and insects last shared a common ancestor 550 million years ago, it is reasonable to conclude that nature really did reinvent the same mechanism for flight for small organisms.

Another strange mode of locomotion that has been on nature's to-do list is the snake's slither. Given a snake's reputation, you might think that nature made the leap from lizard to snake only once. But lizards have lost their limbs, elongated their bodies, and learned to slither at least twenty-six times as life evolved.[13] The process can take tens of millions of years, and it can also reverse itself, restoring digits that had previously been lost from forelimbs.[14]

Sneakers on Lord Howe Island

I saw a striking example of convergent evolution when I visited Lord Howe Island, an idyllic spot some five hundred miles off the coast of Australia. I had seen a picture of the island as a child, and I always thought that a trip there would be the greatest of adventures.

After just a few minutes ashore, my wife and I spotted a lovely butterfly resting on a bush, while another butterfly—evidently the male of the species—flitted around her without landing. The hapless male put on his show for five minutes, ten minutes, perhaps more—how males are humiliated throughout the animal kingdom!—but whatever sign he was looking for from the female, she didn't seem to provide. Eventually, a third butterfly came to visit the now-receptive female. The first butterfly, tired out from his presentation, could only yield his spot to the lucky interloper.

Figure 13.2. White terns on Lord Howe Island, Australia. After a male (*top*) has courted the female, a "sneaker male" (*left*) flies in to oust him. Similar behavior is found in insects, cuttlefish, and many other animals.
Credit: Photograph by the author.

We thought no more about this scene until that evening, when we strolled along a quiet path. There, perched on a tree limb above us, was a lovely white bird and another hapless male flitting around her (Figure 13.2). It was exactly the same scene: a humiliated male flapping wildly and going nowhere, intent on showing the female that yes, he can fly, he's got stamina, he'd make a great mate. The bird was a white tern, endemic to the island and unusual in that the female lays its egg in the barest fold of a tree limb. It balances the egg precariously without benefit of a nest or other support.

We watched for a few minutes. Sure enough, as the male grew tired, an interloper flew right in: another male displacing the first after he had done all the work! When I mentioned to a biologist friend how surprised I was that both butterflies and birds had evolved the same sneaky behaviors, he replied that "sneaker males" are common throughout the animal kingdom.

Perhaps the most unusual example of a sneaker male is among cuttlefish, which are the world's masters of camouflage. When a male cuttlefish is courting, he converts his skin to a manly, colorful display to impress the female. But he does this only on one side of his body: the side that faces the female. On the other side of his body, he sports the coloration of a female. That way, other males don't realize that he's out a-courtin'. Seeing his female-colored side, the other sneaker males might even waste their time trying to impress *him!*

What Is Convergent Evolution Telling Us?

Convergent evolution is one of the most striking features of life, but biologists are still arguing about what it signifies.[15] Do the hundreds of examples of convergence show that life is deterministic, or that natural selection is more powerful than we thought, or that convergence is simply to be expected on the basis of chance? Or do all of these apply?[16]

Until recently, the two areas of biology that bear on these questions had little to say to each other. Evolutionary biologists piece together portraits of life in the large. They painstakingly document patterns of life that stretch over vast spans of time, based on evidence as subtle as faint imprints in ancient rock. By contrast, molecular biologists study life in the here and now. They are faced with the staggering complexity of life at the molecular level, organized in hierarchies of cells, tissues, and organs that form a kind of mini-universe of their own.

Recently, however, thanks to the new tools of biotechnology, these two fields have forged a partnership that yields insights into how nature works its magic. We can see, for example, how life's infrastructure is able to innovate boldly while still preserving what already works. One strategy is to duplicate genes so that the cell has a greater chance of holding on to one good copy, while the other copy has greater freedom to mutate.

A striking example of this is that species of spruce and pine have independently evolved a whole suite of traits that help them adapt to cold environments. (You may think that spruce and pine are very close genetically, but they are not; they have not had a common parent for fifty million years.) A large team of researchers was able to identify a million mutations in twenty-three thousand genes in these trees. They found forty-seven genes related to the cold adaptation that had evolved in parallel. These innovation genes were enriched in genes that had been duplicated.[17]

Another feature of convergent evolution is that it typically involves mutations in "hot spots"—that is, regions of DNA that regulate which genes are turned on and off, as opposed to mutations that change a single enzyme or structural protein. A very small change in a control region can lead to a large change in the organism itself. The organism doesn't have to invent new materials; it just puts the same materials together in new ways. The genes that are the prime targets of convergent evolution are the ones that affect the activity of other genes by controlling when, in what tissues, and by how much a gene is activated.

You *Can* Get There from Here

Life has another extraordinary feature that helps to explain how it can be both conservative and innovative at the same time. First, it turns out that a gene can mutate stepwise, one letter at a time, and yet still preserve its function, even after many such mutations have significantly altered the gene's DNA sequence. There are bacterial genes that preserve their function even after 80 percent of their sequence has been altered! The landscape of evolutionary success appears to be very broad; there are many pathways of mutation that preserve function. Nature is wonderfully redundant.[18]

At the same time, there are single mutations that can completely alter the function of a gene. It is nature's version of the Robert Frost poem in which two roads diverge in a wood, and taking one of them makes all the difference. In this case there are a vast number of ways in which the roads of mutation can diverge; some paths lead to stability, while others lead to radical change.

Some analogies may help. The song "Singing in the Rain" evolved stepwise from an initial melody that didn't quite satisfy the composer. When you listen to the successive drafts of the song as it evolved, it is easy to tell that each version is the same song. Slightly different notes, to be sure, but all in the same key, all with the same upbeat lilt. The same holds true for the evolution of a Beethoven symphony: slightly different notes in each successive draft, but all of the versions are clearly the same symphony.

But if you were to substitute just a few different notes in either of these melodies, you might suddenly hear the melody in a minor key or in a mode (neither major nor minor), and this would change the entire sound of the piece. A few strokes here and there, and suddenly you have a piece that appeals to an entirely different audience.

To see this visually, consider the sentence, "This is what I mean." Mutate a few of the letters and get the sentence,

This iz wut I meen.

The original meaning is still clear. But what if you change the letters in the following way?

Thus if wet I'm mean.

Now the sentence is headed in an entirely different direction.

Life's infrastructure has the same properties. It preserves meaning by being forgiving of change along certain, accessible paths, yet it also fosters innovation at the same time along nearby paths. It walks the tightrope between preservation and innovation.

Story and Script

Convergent evolution is unsettling, because it gives the appearance that life is following some kind of script. Not a script set in stone but the kind that some directors use when they give the players great freedom to improvise. At the level of molecules, nature seems to be guiding, but not dictating. If you look solely at the molecular level, you may miss the larger story—and stories often have their own truths.

I learned about the subtlety of scripts and stories from my wife while we were watching an episode of the television show *ER*, a hospital drama. In the episode, the doctors wheel in a man who has just been in a car crash. "I have a terrible headache," he tells the doctors. (Who wouldn't after a car crash?)

The doctors then order an abdominal CAT scan. (Who wouldn't after a car crash?)

My wife, a family physician, shook her head and murmured, "Pheochromocytoma."

"What?"

"That poor fellow has a pheochromocytoma. It's a rare cancer of the adrenal glands." I thought she had lost her mind. How could she make the leap from a headache to cancer? Sure enough, after the commercial break, the surgeon gingerly approaches the patient and says, "I'm sorry to have to tell you this, but you have a pheochromocytoma."

I am as rational as the next person, but I couldn't see how you could diagnose a rare cancer on the basis of a headache.

"You can't," my wife explained. "But I know how they construct the stories in this show, and that abdominal scan was a giveaway. They were looking for a cancer that causes headaches and high blood pressure. It's good drama."

That's how I see convergent evolution at the moment. It doesn't yet make sense entirely, but what wonderful drama! If you pay attention only to what seems rational, you can miss the subtlety and underlying logic of nature's story.

Understanding convergent evolution at the molecular level has really just begun. At some point, we will be able to decipher the entire history of evolution that is preserved in our genes.

But for the purposes of our story, it is enough to know that convergent evolution exists and is now considered the norm.[19] If life is somehow written into the universe's building plan, we might expect that—when confronted with the same environment—similar forms, functions, and behaviors would keep bubbling up, irrepressibly, from the endless ocean of creation.

And that is what we find.

14

The Sensational Sensations

What is your earliest memory of childhood? Can you describe it now, so many years later? Chances are that your memories are not about deep thoughts or conversations or even feelings. Chances are that you describe your memories as sensations.

To writer James Agee, growing up in Knoxville in the summer of 1915, the memories of childhood included "a streetcar raising its iron moan," "the dry and exalted noise of the locusts from all the air at once." It was the sight of "people in pairs" and "the taste hovering over them of vanilla, strawberry, pasteboard and starched milk." It was the sense of the night as "one blue dew."[1]

Of all life's mysteries, none is deeper than physical sensations. They are not merely our portal to the world. Their existence provides the strongest evidence that life is somehow written into the universe's script and is not a cosmic accident.

You might think that the sensations are anything *but* mysterious. After all, we know the chemistry of vision in exquisite detail, right down to the biochemical changes that molecules in the eye undergo when struck by light. We know the structure and function of the retina, and we can operate on it with impunity. We have traced the neural pathways for vision back to the brain and have mapped the brain's visual cortex. We can model how the eye and brain detect the faintest motions, and we can build machines that do the same.

But that is not the sensation of vision. That is just the hardware—the molecules and cells and tissues that make vision possible. The product of all this hardware is what is mysterious. The sensations are literally in a class by themselves; nothing in the known universe is comparable. To convince you of this, we'll need to do a few experiments. They are very simple, but they require you to see things in a radically new way.

Here is the first experiment: Close one eye and examine your surroundings. Got it? Now open both eyes and examine your surroundings.

What's the difference between the two experiences? How would you describe to someone else the sensation of seeing with both eyes compared to just one? (This isn't a trick question.) You have a wider field of view, of course, with two eyes. But how would you describe *the sensation of depth?*

You'll find it difficult to put the sensation into words. You might say that a more distant object looks farther away, or that you can now see in three dimensions. But you'll find yourself trying to describe the sensation by referring to the sensation itself. Words seem to fail here. You either experience it or you don't.

A machine can also detect distance, of course. Many new cars have a dual camera system that can capture two two-dimensional images of the road ahead and use them to calculate the distance to the car in front of you. But the result is a number, not a sensation.

Try a second experiment: Find a red object in the room and focus on the sensation of red. Now do the same for a blue object. Try to really feel the sensation; put every other association out of your mind.

How would you describe the difference between the sensation red and the sensation blue to someone who had never seen either?

You'll find this difficult to do. In fact, it is impossible to do without invoking the sensation itself. You might say that red is a warm color like fire, while blue is cold, like the color of ice. But if you had not seen the colors of fire and ice, you would have no idea what this meant.

You might be tempted to define the color red by describing the physical conditions that produce it. A certain wavelength of light, under certain conditions, a scientist might say, would lead to the brain state of seeing red. You might then map out the activity of every nerve cell in your brain and say this is the physical state of seeing red. You might then do the same for the color blue. At the end of this study, you would be able to say the brain state X produces the sensation that people call A. And brain state Y produces the sensation people call B.

But here is the catch: This tells us nothing about the sensation of seeing red. It merely assigns a different word to each brain state. More importantly, we have no way to explain or predict why brain state X produces sensation A rather than sensation B. No way, even in principle.

The reason is that there are no scientific descriptors for the sensation of seeing red or blue. For example, these sensations make no contact with the fundamental concepts of physics, such as mass, length, and time. You can experience the sensation, but you cannot characterize it with the parameters of science.

It is truly remarkable that no other aspect of life has this property. For example, there is a structure inside each of your cells called a mitochondrion, which supplies the energy needed to power the cell's activities. If you could somehow specify the location of every atom and molecule in a typical mitochondrion, along with the molecules that enter the structure and leave the structure, then you would have given a full description of the mitochondrion. Even if you had no idea how this structure actually functioned, you could eventually figure it out from an understanding of how molecules interact.

But red and blue? If you hadn't experienced them, no knowledge of chemistry and physics and biology would help you in the least.[2]

Have you ever tried to imagine what entirely new colors might look like? Ones that can't be mixed from the colors we can see? Other animals can see parts of the spectrum that we can't. For example, bees can see ultraviolet light—a wavelength of light just beyond the furthest violet. On the other hand, they can't see red, at the other end of the visible spectrum.

Remarkably, many plants have accommodated to bees' visual world. For example, sunflowers have developed markings on their petals that can only be seen by ultraviolet-sensing insects. These distinctive markings guide the bees toward the nectar, just as runway markings at an airport guide planes to a safe landing.

What color does a bee experience when it sees ultraviolet light? I posed this question many years ago to Béla Julesz on a visit to the Bell Laboratories in New Jersey. The labs are famous for the development of the laser, the transistor, radio astronomy, and the light-sensing chips that revolutionized photography. Julesz was renowned for his work on visual perception, including depth perception. I was young and impressionable at the time, so I vividly remember his response to my naive question.

His eyes were shining, and a childlike grin spread over his face. "Ahhh," he breathed, almost relieved that someone had asked him the question. "We will *never* know what even a *mosquito* sees."

Never is a word that scientists don't like to use, because history has been so unkind to those who use it. (At one time it was thought we would never know what the stars are made of.) Yet Julesz felt confident using the word because he realized that, just as beauty is in the eye of the beholder, so is the experience of sensation. It doesn't make contact with the bedrock of physics, for example.[3]

The sensation of color is an example of an *emergent* phenomenon. This is a phenomenon that arises from a complex system and that cannot be understood simply as the sum of its parts.

A mob, for example, is capable of doing things—burning buildings, wreaking havoc, even killing innocent people—that no one of its members would dream of doing individually. If you have ever started to cross the street while the light is red, you may have noticed that others will automatically follow suit, without even looking for oncoming traffic: emergent behavior. Other examples are the flocking behavior of birds and the schooling of fish.

If you belong to a group, such as a book group, you'll have noticed that it has a dynamic that is different from, say, that of two friends chatting. Even the size of the group is important. Studies have shown, for example, that the best science and the best Broadway musicals are produced by teams of about seven people. And the best group makeup is a mixture of novices and experienced people. The novices contribute out-of-the-box thinking, while the old-timers provide experience and judgment.

But these are generally considered examples of weak emergent behavior, because you can explain some of a group's behavior in terms of the psychology of individuals. The sensations are fundamentally different. They make no apparent connection with the molecules that produce them.

You might be tempted to think that the sensation red, say, *is* the sum of its parts. Suppose you could create an electronic copy of your brain. An exact duplicate that mimicked every connection in your brain while you saw a fire engine pass by. The electronic "you" would detect light, would have an electronic retina, and would process the signals as similarly to your own brain as possible.

Would that electronic brain see fire-engine red?

Perhaps. But how could you ever find out? Even if your carbon copy could speak to you, how would you know what sensation it was experiencing?

More importantly, even if a machine *were* to experience the sensation red, that would tell you nothing. You'd still have no idea why this elaborate setup produces red rather than blue or green. You have an assembly of molecules on the one hand and the amazing sensation of red on the other. What in the world is the connection?

You might wonder why we are discussing the restricted field of sensations rather than the broader arena of consciousness. I don't really know what the word *consciousness* means; it is fuzzy around the edges. It seems to be a conglomeration of all kinds of things, such as emotions, intelligence, and sense of beauty, as well as the sensations. Some of these abilities are already reproducible in a machine; some are not. There are computer programs that write passable music, though you can usually tell that something is amiss. But color, depth, taste—these are crisp; we don't all agree on them, but we do know what they are.

What are the sensations telling us about life and the universe? To some scientists, remarkably little. The physicist George Gamow wrote, "The basic manifestations of life like growth, motion, reproduction, and even thinking depend *entirely* on the complexity of the molecular structures forming living organisms, and can be accounted for, at least in principle, *by the same basic laws of physics* which determine ordinary inorganic processes" (emphasis added).[4]

Gamow's message is, "Don't worry, physics has it under control." You'll note that he doesn't mention the sensations explicitly, perhaps because it is so easy to take them for granted and not even notice them. As far as we know, our senses do not violate the basic laws of physics—but those laws are not sufficient to account for the senses, even in principle. Like many scientists, Gamow wanted to close the door on nonscientific ways of thinking, such as vitalism, which imagines that there is a mysterious life-force that animates living cells and that is independent of atoms and molecules. But in doing so, Gamow also closes the door on one of the most interesting of all scientific mysteries: how awareness arises from mere molecules.

The physicist J. Robert Oppenheimer held the opposite view. "It seems rather unlikely," he told a BBC radio audience, "that we shall ever be able to describe in physicochemical terms the physiological phenomena which accompany conscious thought or sentiment or will."

He went on to say that even if we could one day understand the physical basis of consciousness, "it will not be itself the appropriate description for the thinking man himself, for the clarification of this thoughts, the resolution of his will, or the delight of his eye and mind at works of beauty."[5]

The cosmologist Andrei Linde goes further. He reminds us that we tend to view consciousness the way we used to regard space in Isaac Newton's day. Space was just there, in the background, a convenient tool for describing matter, which at the time was the real subject of interest. But thanks to Albert Einstein, we now know that space has a reality of its own, connected to, wired into, matter and energy.

In the same way, we have considered consciousness to be in the background, just a tool for describing the "real" world out there. "But let us remember," notes Linde, "that our knowledge of the world begins not with matter but with perceptions. I know for sure that my pain exists, my 'green' exists, and my 'sweet' exists. I do not need any proof of their existence, because these events are a part of me; everything else is a theory."[6]

According to Linde, we perceive the world around us to obey certain laws—the laws of physics, for example—and we start to formulate a reality "out there," beyond our perceptions. And so good are our sensations at projecting a world "out there" that it is easy to overlook our own consciousness and thrust it into the background: "Is it not possible that consciousness, like space-time, has its own intrinsic degrees of freedom, and that neglecting these will lead to a description of the universe that is fundamentally incomplete?" Linde asks. "This model of the material world obeying laws of physics is so successful that soon we forget about our starting point and say that matter is the only reality, and perceptions are only helpful for its description. This assumption is almost as natural (and maybe as false) as our previous assumption that space is only a mathematical tool for the description of matter."

"Will it not turn out," Linde continues, "that the study of the universe and the study of consciousness will be inseparably linked, and that ultimate progress in the one will be impossible without progress in the other?"

What a thrilling thought: that universe and consciousness could be inseparably linked, as John Archibald Wheeler too had imagined. Were this to be the case, then our senses and intelligence and awareness and feelings would be much more than mere tools for survival.

They would be the portals by which we come to know the universe itself, to see all the way out in space and all the way back in time, to ask deep questions whose answers we already know might be beyond our grasp. They would be gifts from a universe to which we are inextricably bound.

15

Design without a Designer?

Let's take a brief side trip to see something beautiful and mysterious: the most complex known pattern in the universe. This pattern bears on the question, Can simple rules—for example, the rules with which the universe was born—create something as complex as a human being? Our side trip will raise deep questions to which I don't have the answers, but perhaps you can help.

The pattern we're going to look at is called the Mandelbrot set, named for the French mathematician Benoit Mandelbrot, who did much of his work at Yale University. You may have seen the Mandelbrot set before—or parts of it, since no mortal can explore the whole pattern—but we're going to look at it from a perspective that is different from usual.

The Mandelbrot pattern is similar to the paint-by-number drawings that we made as children, with two big differences. First, in paint-by-number, an artist drew the picture first, then asked us to fill in the colors. The Mandelbrot pattern is determined by a simple mathematical rule. There is no artist at all—unless you consider mathematics to be part of nature, in which case nature is the artist. We'll come back to that statement later, because the question of who is the artist is much more subtle than you might think.

The second difference is that paint-by-number drawings provide big areas for us to color in. They are not subtle drawings. But in the Mandelbrot pattern, the mathematical rule determines a color for every single point on the canvas. The full image is made of an infinite number of colored dots. That means that you could zoom in on the image as much as you like and still see a pattern. And, as you'll see, the pattern is ever varying; it never exactly repeats itself. Drawing this pattern is not something that a human wants to do, because there are simply too many points to color in. A computer is needed to create the image.

What is the mystical mathematical rule that produces the Mandelbrot pattern? I'm not going to say much about it; the math would distract us from our mission. But I'll give a brief overview.

We'll take a canvas and draw two axes on them as though we were going to make a graph. The axes just serve to label every point on the canvas with two numbers. The origin, where the axes intersect, is (0,0), and points farther away are labeled by their distance from the two axes. The important thing is that every point on the canvas is denoted by its two numbers. Points that are near to each other will have very similar numbers.

To determine how to color each point on the canvas, we play a simple mathematical game, using the two coordinates of the point. You can find a version of the rules that uses only arithmetic in the notes.[1] The important point is that the rule for deciding how to color a point is straightforward—and the same rule applies to every point on the canvas.

Figure 15.1. The Mandelbrot set. Points inside the set are shown in black. At the boundary of the set, the pattern is infinitely complex. A portion of the area in the rectangle (indicated by the arrow) is enlarged in each successive image. Credit (Figs. 15.1–15.5): Image created by Wolfgang Beyer using *Ultra Fractal* 3. Wikimedia Commons (CC BY-SA 3.0).

The astounding thing is that a simple rule or procedure can lead to a pattern of infinite complexity. Before we discuss what the pattern might be telling us, let's explore a small portion of it.

A wide-angle view of the Mandelbrot pattern is shown in Figure 15.1. The black shows all the points that are definitely inside the Mandelbrot set, and the gray background shows the points that are definitely outside. From a distance, there is nothing especially striking about the pattern. There is a heart-shaped blob, with smaller appendages budding from it. The whole thing sits in a calm sea of gray.

But strange things appear to be going on at the border between the Mandelbrot set of points and the rest of the canvas. In the real world, most of the interesting action also happens at boundaries. Think of life on Earth, occupying the thin boundary between the inside of our planet and the outer reaches of space. Or think of the world's great cities, poised on a seacoast or riverbank at the boundary of commerce with the world.

If we zoom in on one of these border regions of the Mandelbrot pattern, we see the fantastic pattern in Figure 15.2. (If you choose another border region to explore, you will get a different pattern.)

By the next step, Figure 15.3, it is clear that we are in another world, an Alice-in-Wonderland world, utterly unexpected and delightful. If we zoom in further, we marvel at the sixteenfold symmetry of the seahorse pattern, which appears out of nowhere (Figure 15.4). And there, in the center, we see the Mandelbrot pattern again, only this time much smaller than the original, since we have zoomed in many thousandfold.

There are many possibilities to explore. There are the psychedelic spirals, luring us ever further into smaller and smaller scale. There are curlicues on these spiral bannisters that are worlds of their own, entire universes, really, given that even these tiny regions are infinitely complex as you magnify them.

We'll stop with Figure 15.5, because we can never explore the entirety of this pattern. It is vast beyond reckoning. At this scale, we have magnified the original image so much that it would wrap around the entire Earth! Besides, we have to stop here, because our computer will eventually run out of computing power: it can't

Figure 15.2. Part of the Mandelbrot set, magnified one hundred times. This region is nicknamed the Seahorse Valley, after the graceful spiral shapes that appear.

Figure 15.3. Part of the Mandelbrot set, magnified thirteen thousand times. At the very center, a miniature version of the Mandelbrot shape appears.

Figure 15.4. Part of the Mandelbrot set, magnified one hundred thousand times. This microscopic version is not the same as the original shape. Where did the beautiful symmetry come from?

Figure 15.5. Part of the Mandelbrot set, magnified 180,000,000 times. Resembling the seeds on a sunflower, the spiral beckons us inward to discover an endless succession of patterns.

handle numbers much smaller. But the pattern will continue infinitely far into the infinitesimal—a place we cannot go.

No human designed this pattern, nor can any human explore even a tiny fraction of it.

You'll notice something fantastic about Figure 15.5. We have zoomed in a hundred millionfold, and yet the resolution of the pattern is exactly the same as when we began. There are no atoms or molecules or building blocks of this pattern. It is as fresh and as robust at this tiny scale as it is at the larger scales.

The reason for this is that we have not really expanded the original pattern, as though it were some tangible, printed image that we blew up in the darkroom. Instead, *we have changed the scale of our gaze.* In this strange, mathematical world, a scale of distance that is infinitesimally small is just as valid as the larger scales of distance. We saw this idea in Part 1, where we learned that the universe is not expanding outward at all. Rather, it is *fractalizing,* expanding inward to smaller scales of distance, just as we see here. And we can see echoes of Bernhard Riemann, for whom the scale of distance was malleable, and who wondered whether the world might be very different at the smallest or largest scales of distance.

The Mandelbrot pattern is important to our story of the universe for several reasons. It illustrates that simple rules can give rise to extraordinarily complex patterns. If the Mandelbrot pattern is possible, then it becomes easier to grasp that something as complex as a human being might emerge from the simple rules with which the universe was created. One of the first artists to work with Mandelbrot images, Katherine McGuire, said that she loved these images because they are like us: finite on the outside, yet infinite on the inside.

The Mandelbrot pattern is a true miracle, in the sense that I have been using the word in this book: we understand it, we can write down the mathematical formula that produces it, yet our jaws drop that such a thing is possible.

Now for the most difficult and perplexing questions: Who or what designed the Mandelbrot pattern? Would this pattern exist if we weren't here? Can there be design without a designer?

You might be tempted to say that the pattern is inherent in the simple rules that Mandelbrot formulated. The pattern would exist in some ethereal, ideal mathematical world even if no intelligent beings had been created first. The relationships of mathematics would exist in our universe regardless. They are an ideal to be discovered by inquisitive beings, here or elsewhere in the universe.

The story is much stranger than that, however: the Mandelbrot pattern can be constructed, *but not predicted*. Even if you know in detail what one part of it looks like, you can't predict what an adjoining space will look like. You have to crunch the numbers. Infinitesimally close regions need not be at all similar to each other; a small change in distance can lead to an enormous change in the pattern.[2]

This means that the rule that tells us how to create the Mandelbrot pattern is not the same as the pattern itself. The pattern has to be constructed, brought into existence with actual computers and display media and real people at the helm.

This brings us full circle to John Archibald Wheeler's image of the ultimate equations of physics sitting on a blackboard. When all is said and done, the equations just sit there. So too does Mandelbrot's wonderful rule just sit there on the page. Only because humans have been crunching the numbers do we have even this faint glimpse of what the pattern looks like.

Today, there are several smartphone apps that let you explore the Mandelbrot pattern for yourself. Children and adults all over the world are bringing into existence parts of the pattern that no one has ever seen before and that perhaps never will be seen again. Just as nature somehow gives body and substance to the universe, so too are people giving life to the Mandelbrot pattern, bringing it from the ethereal world of mathematics into the light of day.

Like the universe, it only exists when you construct it. Like the rest of the universe, it is still coming into creation. The Mandelbrot pattern is nature's fantastic paint-by-number drawing, painstakingly drawn by its curious children.

16

"Who's There?"

If you really want to find out whether life is written into the machinery of the universe, there is a simple experiment you can do. Just rewind the history of Earth back to the moment of its birth, let it evolve in time, and see what you get. Run this experiment not just once or twice but over and over again. A thousand times. A million times. How many times do you get life, and under what conditions? Does the life look remotely like the world outside your window? Does anything resemble what you see in the mirror?

The truly wonderful thing about this experiment is that it not only can be done—it has already started! Nature is carrying out this experiment even as you read this. There are estimated to be several billion Earth-size planets in our Milky Way galaxy alone, in various stages of development.[1]

It's not quite the same as rewinding Earth 1.0, but it's good enough. And if we can look for signs of life on even a tiny fraction of those worlds, the experiment may tell us what we want to know about life's place in the universe. Technologically, looking for life is extremely challenging. But it holds the promise of answering John Archibald Wheeler's question.

Awakening

When most of us were born, the nine planets that orbit our Sun were the only planets known in the observable universe. As children, we dutifully learned the names of these worlds, their distances from the Sun, and what they were like. We were thankful to live on a planet so welcoming to life, but we also marveled at the strange and alien landscapes on the other eight worlds in our solar system. I still remember an eerie diorama at the local planetarium, glowing under ultraviolet light, showing an artist's conception of Pluto, the farthest outpost among the planets.[2]

Everyone assumed that some of the stars out there in the night sky must be orbited by planets as well, but it was just speculation. A famous equation by Frank Drake attempted to bring order to the speculation, but it was really just a string of questions to which there were not yet good answers. "What fraction of stars might have planets? What fraction of those planets might orbit in the habitable zone of their star, where water remains liquid?" And so forth. It didn't give us answers, but it gave many of us hope that one day there might *be* answers.

And then, one day in the 1990s, astronomers rather quietly detected the first planet orbiting another star. Then another, and another. At first it was a trickle. These early discoveries turned out to be large, gaseous planets like Jupiter, but unlike Jupiter they orbited close to their stars, zipping around in a few days rather than the years it takes Jupiter to make one circuit around our Sun. And unlike Jupiter, these were hot worlds, so close to their stars that their atmospheres puffed up, like fowl strutting their mating plumage.

They were not really interesting worlds to the layman, but they did hold an interesting lesson: there is nothing like them in our solar system. They were not what astronomers expected to find.

These worlds defied the theories of planet formation. How could a Jupiter-size ball of gas form so close to a star without boiling away? Perhaps the planet formed somewhere cold, far from its star, then migrated in? Astrophysicists went back to their drawing boards to model new theories.

These alien worlds are called exoplanets. They aren't given fanciful, mythical names like the planets in our own solar system, merely designations rooted in the stars they orbit: HATP-3b and TRES-3b and a whole alphabet soup of other acronyms that astronomers use.[3] The names were uninspiring, but it was the little *b* that sent chills up and down our collective spine. The *b* after the name of the star indicated that this was a planet—a companion to the star, a world, a mystery. There was hope that the star might have other planets: a *c* and a *d* and maybe even an *e*. It gave us a shiver to think that our own planet, the one with six thousand names, including Earth, could be some alien civilization's *d*—the third planet from our star. A tiny letter to represent a whole world.

And then the floodgates opened. There were a dozen exoplanets known. Then dozens. Then a hundred. And then NASA sent into space the Kepler probe, designed to stare at a patch of stars without interference from Earth's turbulent atmosphere. For years, Kepler stared at stars in the direction of the constellation Cygnus, the Swan. That representative patch of the sky contained one hundred thousand stars. Kepler watched for the periodic dimming of each star, a faint decrease in brightness that might indicate a planet was eclipsing its star and blocking some of its starlight.

The conclusion: our Milky Way galaxy is teeming with planets.

Thousands of exoplanets have already been detected. The discovery of these worlds has upended existing theories about how solar systems form and overturned our ideas about what planets

should be like.[4] The discoveries to date show that nature's repertory of worlds is far more varied than we had imagined. These
worlds are not like our own!

There are planets that are darker than any world known, darker
even than coal. There are ocean worlds made almost entirely of
water and steam. There is a giant planet so infernally hot that it
rains glass at higher altitudes and rains iron down below.

There are planets that orbit two stars at once, treating a would-
be visitor to two sunrises and sunsets every day. There are planets
that zoom around their suns so fast that their "year" is compressed
into a few days. There are planets whose looping orbits take them
alternately nearer and farther from their suns, creating an endless
cycle of freezing and thawing.

There are planets being observed in the long process of formation, and planets orbiting stars that have already imploded, and
planets that are disintegrating as they fall into their stars.

And then there are the planets that are the size of Earth, maybe
a little larger, maybe a little smaller.

There are the planets that orbit peacefully in the so-called habitable zone of their star—neither so close to the star that water
boils away nor so far that water freezes to ice. Right in the middle,
where water is the liquid we know and love and depend on.

Based on the sample of planets that have already been detected,
*there are estimated to be several billion Earth-size worlds in the habitable
zones of their stars, throughout our Milky Way galaxy.*[5]

The Challenge

Imagine that you are an alien who has traveled across the loneli-
ness of interstellar space and has just now arrived at our solar

system. You have passed the orbits of Pluto, Neptune, and Uranus and have just arrived at the planet Saturn. Your view might be similar to the vista in Figure 16.1, taken by NASA's Cassini space probe at Saturn.

You are impressed by Saturn's magnificent rings, but you spy what you think might be still another planet: Earth appears as just a dot, to the left of the rings in Figure 16.1. Earth is still about nine hundred million miles away in this image, as is the Sun, which is hiding behind Saturn. Our solar system is huge. It took the NASA spacecraft seven years to get here from Earth.

That dot is a whole world. At the moment the image was taken, my family and friends happened to be on vacation at a lake in western Massachusetts. You are on that dot as well, along with every other living thing on Earth. As you can see, the photo doesn't do any of us justice.

In fact, from just this image, an alien would not be able to tell that there was *any* life on the planet.

Now here's the challenge: The nearest solar system to our own is more than a million times farther away than Earth and the Sun are in this photo. That means that the exoplanet will appear a million times smaller than even this dot. It will appear a million times closer to its star—so close that the star and exoplanet would appear as a single speck of light. Furthermore, the star and exoplanet would appear a trillion times fainter![6]

You might think it would be hopeless to detect exoplanets, let alone describe anything about them. On the contrary, it turns out to be easy to detect them. The simplest method is an indirect one. You simply look for the periodic dimming of a star's light as one of its planets passes in front of the star, blocking some of its light. (This "transit method" works only for a planet whose orbit is ori-

Figure 16.1. *Top:* A photograph of you, plus Saturn and its rings. *Bottom:* An expanded view of the rectangle in the top photograph. The small dot to the left of center is Earth, about nine hundred million miles in the distance. Even the nearest exoplanets are a million times farther away than this dot. Credit: NASA / Cassini / JPL / Space Science Institute.

ented edge on as seen from Earth, but there are many such planets.) From the dimming, it's possible to figure out the size of the planet, how far it is from its star, and other rudiments of a planetary portrait.

In less than three centuries, civilization has progressed from observing the transit of the planet Venus, within our own solar system, to observing thousands of transits of distant alien worlds as they eclipse their own stars. Suddenly, the universe seems to be at our fingertips.

Remarkably, you don't need the powerful Hubble Space Telescope to detect these planets by the transit method. You can do it with the telescope that Santa Claus gave you for Christmas, at least for the largest planets. It's no surprise that the first planets to be detected this way were big and hot. A big planet blocks a lot of its star's light, so the dimming is pronounced and easier to detect than a fainter signal. And a planet that is close to its star is easier to find, because it orbits faster than planets on the outskirts of their solar systems. Faster means there are more transits in a given time, so more chances to detect a signal. Earth, for example, orbits the Sun once a year, so if you want to detect three transits—the minimum for confirmation—you have to observe for years at a time.

Life Closer to Home?

As our alien visitor leaves Saturn and heads past Jupiter, it can't help but notice the brilliant moons of Jupiter. One of them is particularly intriguing. Europa is one of the four moons of Jupiter that Galileo discovered through his telescope just over four hundred years ago. Europa is about the same size as our own Moon, and both moons are about the same distance from their respective planets. That's where the similarity ends.

Europa's icy surface is carved with cracks containing a strange, reddish-brown material of uncertain composition. Europa is so far from the Sun that the temperature there is −260 degrees Fahrenheit—and that's on a good day, at the equator. At the poles, the temperature is a hundred degrees colder.

Who would have thought that this unpromising place not only has liquid water but also an ocean containing twice as much water as all the seas on Earth! How is this possible?

My first glimpse of the strange story came many years ago, when I went to New York to interview the science fiction writer Arthur C. Clarke. I was fresh out of the biophysics research laboratory and was making a documentary for public television on artificial intelligence. Clarke was the author of the novel *2001: A Space Odyssey*, featuring the intelligent computer HAL (a name cleverly derived from "IBM" by shifting each letter). The novel was based on earlier stories and on the screenplay of the blockbuster sci-fi film that Clarke wrote with Stanley Kubrick.

Clarke was renowned as a futurist, yet on this encounter he seemed to be looking back rather than forward. He insisted on being filmed in New York's old Chelsea Hotel, in the very room where he had penned his best work. Earlier generations of writers had made the Chelsea famous, but now the hotel had been repurposed as a single-room-occupancy structure, meaning that it was populated with tenants who led lives of despair. Throughout the filming, there were assorted cries and moans filtering through the walls. To the film crew's horror, the only ambient light in the room was a bare light bulb swinging from the ceiling. Clarke insisted on being filmed with the same model of typewriter he had used when he wrote his novel; we dutifully found one in a nearby pawnshop.

In Clarke's novel, the adventurers go to Saturn, but the film version substituted the planet Jupiter instead, for various technical reasons. Clarke kept current with developments in science, so he became aware of the intriguing theory that Europa might have an ocean beneath its frozen surface. Europa is heated from within by ordinary friction: the moon is alternately stretched and squeezed as it orbits the massive planet Jupiter.[7] You can re-create this heating effect yourself: just bend a wire coat hanger back and forth a few times, then carefully touch the flex point. Hot!

In Clarke's sequel, called *2010: Odyssey Two,* the author imagines life in an ocean under the Europan ice. And the spaceship crew receive an urgent message from an unknown entity:

ALL THESE WORLDS ARE YOURS—EXCEPT EUROPA.
ATTEMPT NO LANDINGS THERE.

As you might imagine, disaster ensues.

NASA used an excerpt from Clarke's popular fantasy in a foreword to its mission plan to Jupiter and Europa. And, as luck would have it, life imitated art. Not that NASA found life on Europa—far from it—but it found evidence for the ocean that had been predicted years earlier.

The clincher came from Margaret Kivelson's lab at UCLA. An expert on magnetic fields, Kivelson showed that Europa's effect on Jupiter's magnetic field could be nicely explained by the presence of a saltwater ocean sloshing beneath the thick surface ice. Current estimates suggest that the ocean is sixty miles or more deep, much deeper than any ocean on Earth.

And the best part, should you wish to dive there, is that the ocean is roughly the same temperature as the waters of Bermuda. Of course, to explore the ocean, you would first have to drill through at least a mile of ice, and probably much more than that.

Plumes

It may not be necessary to drill through the ice of Europa to find evidence for life. It appears that material from the ocean periodically makes its way to the surface, oozing through cracks in the shifting ice. And reports of a geyser on Europa suggest that it may be possible to sample the ocean water from above the surface.

That is the case with one of Saturn's moons, called Enceladus, which has an ice-covered ocean. NASA's Cassini mission spotted more than a hundred geysers on this tiny moon. The geysers are mostly water, with trace evidence of organic material (carbon compounds). The water has percolated through the hot rocks in the moon's interior—just the kind of environment recently thought to create some of the molecules necessary for life.[8]

Here on Earth, biologists have turned their attention to these undersea hot spots—vents that harbor many strange forms of life. The ancient microorganisms found here, called Archaea, form a kingdom distinct from bacteria and other forms of life. These cells have a biochemistry that is somewhat more primitive than other cells. They thrive in extreme environments, such as the hot pools of Yellowstone National Park, or in extremely salty environments where nothing else could survive.

It isn't likely that there are fishlike creatures swimming in Europa's vast subsurface ocean. For one thing, the dark and airless ocean probably could not supply the energy needed to support the larger forms of life. But microorganisms are certainly a possibility. Their discovery would be a monumental find.

NASA will send a mission to Europa in the 2020s to confirm whether this world has the liquid water, organic molecules, and energy sources needed for life. Europa's surprising ocean shows the importance of being open-minded about what constitutes a habitable zone for life. Just a few decades ago, few would have imagined that liquid water would be found anywhere near Jupiter and Saturn.

Looking for Life on Exoplanets

It is one of the great wonders of science that a single dot of light can be teased apart into its component colors, analyzed, and deciphered to create a portrait of the world from which it came.[9] It is possible to tell, from a speck of light, whether a planet has continents, oceans, an atmosphere, a temperate climate, and even some form of life. Light is truly a winged messenger, bringing us news of regions we could not hope to visit in person.[10]

There are at least three ways that astronomers will be looking for life among the exoplanets. One is to examine the composition of an exoplanet's atmosphere. A potential sign of life is oxygen. On Earth-like planets, oxygen would indicate the possible presence of vegetation. An even better sign is to find both oxygen and methane in the same atmosphere. (Methane is the gas in a kitchen stove, and it is produced by various species of microorganisms.) The oxygen molecule is so reactive that it would quickly combine with methane. To find both molecules in the same atmosphere would suggest that they are continually being replenished, presumably by living things.

Ozone is another form of oxygen that would provide some evidence for life. So is air pollution, for that matter. Some of the exotic molecules produced by industry are not only signs of life; they are signs of intelligent life.

A second approach is to determine the color of the planet. Considering that vegetation has graced the planet Earth for several billion years, it is not unreasonable to imagine that it might also be found on many other habitable words.

Researchers have lovingly gathered the spectra of hundreds of living organisms whose colors might flag the presence of life on another planet. There are purple organisms that fill whole lakes, and shocking pink ones that grow only in salt lakes, and every shade of yellow and green. The chances of finding these specific organisms on another world are probably nil, but I find it deeply moving to see great libraries of data being assembled in the hopes that they can be matched to whatever we do find on another world.

Plants reflect light with a very distinctive pattern. They reflect green light, of course (that's why they look green). But they also reflect infrared light very strongly—so much so that if your eyes could detect infrared light, all plants would look as bright as snow. Detecting this reflected light signature from another planet would be evidence of vegetation.

A third approach is to look for seasonal changes in the color of a planet. If you fly over Vermont or Colorado this summer, you'll see the broad swaths of green forests; but come back in the fall, and you'll be treated to a riot of reds, oranges, and golds. Living things transform the look of a planet, and they do it on schedule.

All of these techniques require large amounts of light to be spread into a spectrum of wavelengths and analyzed. That means patiently gathering light with large telescopes for extended periods of time. NASA's giant James Webb Space Telescope (the successor to the Hubble Space Telescope) may well provide the first evidence of life on another world.

Dwarf Stars and Their Planets

One strategy in the search for life focuses on stars known as M-dwarfs. They sound exotic, but they compose three-quarters of all the stars in our galaxy. They are smaller, redder, and dimmer than our Sun. And because an M-dwarf star is cooler than the Sun, its habitable zone lies closer to the star itself. This makes it easier for astronomers to detect Earth-size planets—not least because an alien "year" for one of these planets would be measured in days rather than the full year it takes Earth to complete one orbit.

An example of a dwarf star is TRAPPIST-1, in the constellation Aquarius. The star is orbited by at least seven Earth-size planets. At least three of those planets are in the habitable zone of their star, where liquid water and life are possibilities.

There is not yet consensus about whether planets around dwarf stars are hospitable to life. One problem is that the surface of a small star lies closer to its core, where the nuclear fires burn. The effect is to make the star's surface much more turbulent and unpredictable than is the case with our Sun's surface. Planets that are close to a dwarf star are bombarded by violent flares and bursts of ultraviolet light from the star. The concern is that these violent events would sterilize orbiting planets and strip away their atmospheres. But the recent discovery of a planet that does have an atmosphere considerably brightens the prospects for life.

Furthermore, close-in planets will have one side that perpetually faces the star, just as one side of the Moon perpetually faces Earth. If you live on the sunny side of the planet, expect much hotter temperatures than are found on the dark side. There is some evidence that the atmospheres of such exoplanets redistribute heat from one side of the planet to the other, again helping the prospects for life.

Many mechanisms have been found which might boost the chances for life on the exoplanets of dwarf stars. For example, ultraviolet light from our star is also a problem right here on Earth, since it damages the DNA in our skin. (Earth's ozone layer absorbs some but not all of the Sun's ultraviolet light.) People can just wear hats or don suntan lotion, but of course animals cannot.

Corals are particularly sensitive to ultraviolet light. Fortunately, many of them have evolved to fluoresce brilliant colors when bathed in ultraviolet. This clever mechanism essentially transforms the damaging energy of ultraviolet into visible light, which is then radiated away from the coral.

Some researchers have speculated that life on planets near M-dwarf stars could have evolved to do the same. In that case, we might want to look at telltale signs of fluorescence from these planets.

The Earth in Baby Clothes

We know that our galaxy is teeming with Earth-size planets. But which Earth do they resemble? Are they like the present-day Earth, bustling with life and with the *signs* of life? Are they like the early Earth, when there was not yet oxygen in the air and the land was a silent desert? Are they like the "snowball Earth," when our planet was a slush ball and the oceans nearly froze?

We will be treated to all of these views as we begin to describe in detail our planetary cousins out there. We will be able to see— at least in principle—how planets like Earth age. We will be able to see Earths in their baby clothes.

If you want to know how a child ages into adulthood, you could of course follow the child for seventy years or so. But it would be better to observe a whole population of people of different ages and draw conclusions from them. This is how astronomers have determined how stars age. They don't wait a billion years for a star to age; instead, they observe billions of stars in various stages of their lives. They make inferences from these observations.

In the same way, we now have the opportunity to see our own planet in an entirely different way. We will see its evolution dispassionately, from a distance.

And, of course, we will at least begin to make observations of nature's grand experiment: Is there life out there? Where is it, and why *there?*

"You Never Call . . . "

If our galaxy is teeming with Earth-size planets, and if many of these planets could host life, then where is that life? Where is the intelligent life? Why haven't intelligent life-forms visited Earth or made their presence known to the rest of us?

This conundrum is affectionately known as the Fermi paradox, after the physicist Enrico Fermi, who helped popularize the question. (Other scientists did a lot of work on the problem, but if you want to take a question about extraterrestrials seriously, then linking it to a famous physicist helps make the question respectable.) The Fermi paradox was once simply the realm of speculation and science fiction. But now that we are in a position to actually gather evidence, the problem looms large.

There are many reasons why we might not have seen obvious signs of intelligent life. One possibility, of course, is that life is extremely rare. Many people don't like that idea. They point out that the observable universe is so large that surely there must be other intelligent life-forms, including some much more advanced than we are. But size has nothing to do with it. The universe was once smaller than you are; as we have seen, it has been expanding continuously. Since it takes billions of years for our kind of life to emerge, the universe must be billions of years *large*. So even if nature only "wanted" one intelligent life-form per galaxy, or even one per observable universe, we would still emerge from the long slumber of evolution to find ourselves in an immense vault of stars. We would still wonder, Where has everyone in the universe gone?

Personally, I find this explanation unconvincing, for many of the reasons presented in this book. Foremost among them is the utterly unexpected robustness of life and its infrastructure over billions of years. Life really is written into the behavior of molecules—and Earth doesn't have a patent on those molecules.

A second possibility is that we are simply precocious. Early arrivers. After all, it takes several generations of stars to create a solar system like ours, and billions of years to produce awareness of the universe. It's not clear whether this process could have been speeded up. Someone has to be first to be aware of the universe. Why not us?

Only a fraction of the stars that will grace the Milky Way have already formed, so most of the opportunities for life lie in the future. Furthermore, the most long-lived stars are the smaller ones; some dwarf stars can live for a trillion years. So, statistically, a species that awakens to the universe would most likely find itself in the future, not our present, where the action hasn't really begun yet. And life should find itself around dwarf stars, which are more numerous and long-lived than our Sun. According to this reasoning, our mere presence this early in the game is a statistical oddity.[11]

There are other possibilities, of course: Perhaps the universe is teeming with intelligent beings, but they have nothing to say to us. Or they have already been here, done that. Or they are communicating in channels that we don't yet understand because we are so primitive.

The most disturbing possibility is one that few of us would want to dwell on, though perhaps we should. The possibility is very real that technological civilizations simply don't survive very long. The mosquito has been here for tens of millions of years; it has proved its longevity. Humans, not so long. Still, one can easily imagine exo-civilizations that have tamed themselves and become long-lived. Where are *they*?

We don't yet know the results of nature's grand experiments with life elsewhere in the universe. But there is every reason to believe that within our lifetimes, we'll hear the historic announcement of the first evidence for life on another world. And with time, we may even discover the answer to John Archibald Wheeler's question. We may finally learn whether life flourishes throughout the universe, or whether our world is one of only a lucky few on which the universe has smiled.

Epilogue:
What Is Worthy of Our Wonder?

On a sun-drenched hillside in the Galilee, in the town of Tzip-pori in present-day Israel, stand the remains of an ancient meeting house and place of worship built in the sixth century. Outside, a wide road installed by the Romans glistens in the sunlight; you can still see the ruts dug into the stone by the ancient traffic of oxcarts. Inside, on the floor, a tile mosaic depicts the constellations of the zodiac, with inscriptions in Greek and Aramaic. At first, everything seems as it should be; you can still make out the Gemini Twins, and Scorpio, and Cancer the Crab. But there is something strange about this mosaic.

The center of the zodiac shows the chariot of Helios, the Greek god who ferried the Sun across the sky. Yet neither Helios nor his mischievous son Phaeton is anywhere to be seen. Instead, the Sun itself drives the chariot. Evidently, the ancient artist decided to forgo the myth and take one small step forward in depicting the natural world. A baby step.

Today, nearly two thousand years later, we have taken the mea-sure of the Sun, the stars, and the galaxies of stars that light the universe. We finally know why stars twinkle and shine; we have dissected them to their core from afar. We can monitor starquakes in suns that are a hundred trillion miles away, and detect their ringing and vibrations almost as clearly as if they were local church bells. And yet science has come so far, and so fast, that it seems story has not had a chance to catch up.

I first realized this years ago when I was doing graduate work in chemical physics. A young freshman mentioned that the chemical elements are forged inside stars and that we are all star stuff. It was a delightful story, but I realized with shame that I had never thought to ask where the elements came from in the first place. To determine the limits of my shame, I surveyed a dozen graduate students in chemistry and asked them, "Where do the chemical elements come from?" Not a single student had ever thought to ask the question. And that meant that their chemistry professors, from their various undergraduate colleges, had probably never wondered either.

None of us are born knowing anything about chemistry. But here we were, young students prepared to devote our lives to the chemical elements, and we had never thought to inquire about their origins. That is like getting married without ever having asked where your spouse came from!

Science has been parsing nature into ever-narrower specialties. So perhaps it is not surprising that our vast edifice of science does not automatically provide us with a sense of place in the universe. We have to actively look for it.

What do you make of this universe? How do you see your place in it? What, if anything, do you think it is set up to accomplish?

Personally, I am struck by the contrasts and contradictions of the universe's story. On one hand, there is the drama of the Big Bang, when matter made its grand entrance. How seductive it is to think of the Big Bang as the creation of the universe. And yet we know that our universe is still in creation. Space, time, and energy are still flowing into existence from nature's seemingly inexhaustible spring.

We think of our universe as immense: more than a hundred billion stars and planets compose our Milky Way galaxy alone; a hundred billion other galaxies are scattered across space, all the way to the boundaries of observation. And yet everything that now exists was once confined to a tiny and extraordinarily dense patch of space, the ultimate brotherhood of creation and equality, every particle separated at birth and sent flying off to regions unknown to develop in its own way.

It is a dark universe, a perpetual night. Yet every thimbleful of space is filled with light of one wavelength or another, traveling in every possible direction. And though it is a look-but-don't-touch universe, where virtually everything is beyond our reach, light is the messenger that brings to us stories from all of creation. It didn't have to be that way.

It is a clever universe. Starting with just the simplest of materials, it has been systematically creating all the structures we see today. It has expanded at just the right rate—not too quickly and not too slowly—to guarantee that stars would form.

It is a universe that delegates. It has handed over the mantle of creation to the stars, which it produces in such profusion and variety that stars can assume all the roles needed to nurture life. It produces massive, short-lived stars that manufacture the chemical elements that will ultimately be assembled into planets and life. It produces smaller, long-lived stars, the sentinels like the Sun, which warm their planets and life as faithfully as a thermostat with a billion-year warranty.

The stars, in turn, have handed over the mantle of creation to the machinery of the living cell. Beyond all reasonable expectation, this machinery of life has endured virtually unchanged for billions of years—more hardy by far than the Rocky Mountains. Yet every living thing it produces is so fragile that life is never far from oblivion. The machinery of life preserves itself, yet it also innovates with such abandon that it has produced millions of species, filling virtually every niche on our planet. The machinery of life is like a printing press that has never given way to new technology yet has always kept up with the times: it has published life's latest editions for three billion years.

It is a buffeted universe, endlessly beset by chance. The decay of a single radioactive atom can rearrange a person's DNA with consequences that will be passed on for a million years. The impact of a single asteroid that just happens to cross Earth's path can wipe out a million species forever. Yet, despite the assaults of time and chance, life has always found a way. Similar life forms develop again and again when life is confronted with the same environment, as though the universe were a jazz musician who could improvise endlessly on just a handful of familiar tunes.

It is a terse universe, succinct. Its building plan was born eons ago, yet the plan is still at work in every cell of the human body, dictating the elegant curve of a DNA strand, instructing a protein how to fold, guiding the reactions of every molecule according to the rules of chemistry. And though the building plan acts without awareness, what an impressive result!

And what an unexpected development: life has handed over the mantle of creation to us. With our species, and perhaps with many others across the galaxy, the universe has created creativity itself. We have already assembled in the laboratory forms of life that could never have existed without us. We have imagined universes forever beyond our reach and argued their plausibility. We have loved, and we have written poems about it.

We do not yet have a complete language for describing or thinking about the universe, and we find that the old words won't do. We think of purpose or intent as being an emergent property reserved for living things. We can't imagine that a hydrogen atom might have a purpose of any kind. And that is right: a cupful of hydrogen does not have a purpose. But a universe filled with hydrogen is a different matter. Animated by its own forces, it has a forward motion. It has a story. And it behaves *as though* it has a purpose.

To me, that is the most puzzling part of the universe's story: it behaves as though it has a purpose, even though that seems to make no sense. As John Archibald Wheeler wondered, how can the infant universe have had any inkling of its future? If I had to guess, I would venture that the resolution of the paradox will involve the nature of time itself. Although time is a physical quantity—it's there in the equations of physics—the *perception* of time is biological. As Wheeler pointed out, the past has no existence except in the records of the present. And as Albert Einstein suggested to the widow of a good friend, the past, present, and future are one and the same to a physicist. The universe seems to be born of a piece; it has integrity.

Are we significant in this story of the universe? It is fashionable to think not. I am baffled by this. *Significant,* after all, means "noteworthy."

I suggest that we are noteworthy. We are the universe's local biographers—or, if you prefer, its press agents. Back in the Dark Ages, when we thought we were special, when we hid our ignorance with myths and legends, that was precisely the time when we were least significant to the universe. But with every advance in our understanding, we move incrementally closer to the universe itself. Who else will tell the universe's story in this part of the cosmos? The universe is looking at us, and we are staring right back, taking notes.

So, as far as we know, we are the universe's most interesting product. Badly behaved? Yes. Disrespectful of other life-forms? Alas, yes. But if you invited the universe to a dinner party, do you really think it would say to the guests, "Hello. I created Pluto and mosquitoes"? No, it would more likely look your guests up and down and say, "Hey! I created *you!*"

We may in some way be fundamental to the universe's existence, as Wheeler and Andrei Linde suggested. After all, everything that we know about the universe comes to us through our own perceptions and sensations, through our own awareness and observation and reason, through our capacity for memory, through our very human sense of time and the flow of history. We have constructed the universe as surely as it has constructed us. But it would be enough just to be an integral part of the universe's story.

A few years ago, the Museum of Fine Arts in Boston displayed a work by two British artists, Tim Noble and Sue Webster. I don't know whether the work is well known in art circles, but I found it one of the most unsettling pieces I have ever seen. In a way, it is a miniature of the universe's story.

You enter a dark room, and there on the floor is a heap of discarded tin cans, pieces of metal, old planks of wood, some string, all connected in various ways. The heap is illuminated by a single floor lamp. It isn't pretty. It looks like what it is: a pile of junk. At first, you wonder, why is this here? And then you see the shadow that it casts on the wall.

It is the perfect silhouette of two people on a park bench, engaged in a lively discussion. There on the wall are all the things that we take for granted: the drape of a shirt, the loop of a shoelace, the gesture of an outstretched hand. It takes your breath away. How in the world is this work of art possible? How can a pile of odds and ends create such a magnificent shadow of *us,* frozen in time?

I feel the same way about you and me, and all the animals we call companions, and the plants that adorn our windowsills, and the wild things we haven't yet met or come to understand, and the story of how we got here, still shrouded in mystery: strange and moving. How in the world is it all possible? I can't imagine how this cosmos could have been put together so cleverly, from so little.

Yet here we are, in a vast, dark room that extends far beyond our telescopes' reach, beyond our reckoning. Up close, in the harsh light of our own Sun and our own gaze, we too are just a pile of nature's building blocks, strung together atom by atom, molecule by molecule, each obeying the only guidance that space, time, and matter provide: a skimpy set of rules with which the universe was born. When you look too closely, nothing exceptional.

And yet, we cast such a long shadow across this wonderful universe.

Notes

Introduction

1. Charles Darwin, "To Henry Fawcett, 18 September [1861]," letter no. 3257, Darwin Correspondence Project, accessed May 2, 2017, http://www.darwinproject.ac.uk/DCP-LETT-3257.
2. John C. Hess, "French Nobel Biologist Says World Based on Chance Leaves Man Free to Choose His Own Ethical Values," *New York Times,* March 15, 1971, 6.
3. John Archibald Wheeler, "Hermann Weyl and the Unity of Knowledge," *American Scientist* 74 (1986): 372.
4. Jacques Monod, *Chance and Necessity: An Essay on the Natural Philosophy of Modern Biology* (New York: Vintage Books, 1971), 145, 171–172.
5. Stephen Jay Gould, "Extemporaneous Comments of Evolutionary Hopes and Realities," in *Darwin's Legacy, Nobel Conference XVIII,* ed. Charles L. Hamrum (San Francisco: Harper and Row, 1983), 101–102.
6. Steven Weinberg, *The First Three Minutes* (New York: Bantam, 1977), 143–144.
7. Quoted in Tim Folger, "Does the Universe Exist If We're Not Looking?," *Discover,* June 2002, http://discovermagazine.com/2002/jun/featuniverse.
8. Alan Lightman, *The Accidental Universe* (New York: Vintage, 2014).

9. Tim Maudlin, "The Calibrated Cosmos," *Aeon,* November 12, 2013, https://aeon.co/essays/why-does-our-universe-appear-specially -made-for-us. See also his *Philosophy of Physics: Space and Time* (Princeton, NJ: Princeton University Press, 2012).

10. George F. R. Ellis, "Does the Multiverse Really Exist?," *Scientific American* 305, no. 2 (August 2011): 38–43.

11. Christian de Duve, *Vital Dust: Life as a Cosmic Imperative* (New York: Basic Books, 1995). See also Christian de Duve, *Life Evolving: Molecules, Mind, and Meaning* (New York: Oxford University Press, 2002).

12. Wheeler, "Hermann Weyl," 372.

13. Ibid.

1. What Is the Universe—and How Large Is It?

1. Quoted in Carl B. Boyer and Uta C. Merzbach, *A History of Mathematics* (Hoboken, NJ: Wiley and Sons, 2011).

2. This odd pattern arises because the plane of Venus's orbit is slightly tilted with respect to Earth's orbit.

3. Edmond Halley, "On the Visible Conjunction of the Inferior Planets with the Sun," in *Abridged Transactions of the Royal Society* (1691), 457–458.

4. Quoted in David Sellers, *The Transit of Venus: The Quest to Find the True Distance of the Sun* (Leeds: Magavelda, 2001), 139.

5. Guillaume Le Gentil, *Voyages dans les mers d'Indes,* translated and quoted by Helen Sawyer Hogg in "Le Gentil and the Transits of Venus, 1761 and 1769," Out of Old Books, *Journal of the Royal Astronomical Society of Canada* 45 (1951): 90.

6. Joseph Banks, *The Endeavour Journal of Sir Joseph Banks,* entry for September 19, 1768, accessed May 20, 2017, http://gutenberg.net.au /ebooks05/0501141h.html.

7. James Cook and Charles Green, "Observations Made, by Appointment of the Royal Society, at King George's Island in the South Sea; By Mr. Charles Green, Formerly Assistant at the Royal Observatory at Greenwich, and Lieut. James Cook, of His Majesty's

Ship the Endeavour," *Philosophical Transactions of the Royal Society* 61 (1771): 411.

8. Michael A. Hoskin, "The 'Great Debate': What Really Happened," *Journal for the History of Astronomy* 7 (1976): 169–182.

9. Virginia Trimble, "The 1920 Shapley-Curtis Discussion: Background, Issues, and Outcome" (prepared for the Seventy-Fifth Anniversary Astronomical Debate, Washington, DC, April 1995, and for publication in *Publications of the Astronomical Society of the Pacific*).

10. Abbot to Hale, January 20, 1920 (Hale microfilm), quoted in Hoskin, " 'Great Debate,' " 170.

11. Ibid.

12. Harlow Shapley, "The Scale of the Universe, Part I," *Bulletin of the National Research Council* 2, no. 11 (May 1921): 171, http://apod.nasa .gov/debate/1920/cs_nrc.html.

13. Ibid.

14. Ibid.

15. Robert Aitken to E. Barnard, January 5, 1921, quoted in Hoskin, " 'Great Debate,' " 174.

16. Heber D. Curtis, "The Scale of the Universe, Part II," *Bulletin of the National Research Council* 2, no. 11 (May 1921): 194.

17. Shapley, "Scale of the Universe."

2. Galaxies Misbehave

1. The method he used is based on the same principle that is used today to measure the speed of a tennis player's serve or a pitcher's fastball.

2. Vesto Slipher, "The Radial Velocity of the Andromeda Nebula," *Lowell Observatory Bulletin* 58 (1913): 58.

3. Edwin Hubble, "A Relation between Distance and Radial Velocity among Extra-galactic Nebulae," *Proceedings of the National Academy of Sciences* 15 (1929): 168–173.

4. Steven Weinberg, *The First Three Minutes* (New York: Bantam, 1977), 30, 36.

5. Edwin Hubble, "A Clue to the Structure of the Universe," *Astronomical Society of the Pacific Leaflets* 1 (1929): 93. Hubble found the idea of an expanding universe so disturbing that nearly a decade later he was still looking for other explanations for why light from the galaxies lost energy. Today, we would consider the expanding universe one of the greatest of all discoveries, but at the time, it was so new and so unreasonable that Hubble was hesitant to promote his own discovery.

3. What's the Big Idea?

1. William Shakespeare, *As You Like It,* 3.2.
2. Steven Shapin, "Of Gods and Kings: Natural Philosophy and Politics in the Leibniz–Clarke Disputes," *Isis* 72 (1981): 187–215.
3. Herman Erlichson, "The Leibniz–Clarke Controversy: Absolute versus Relative Space and Time," *American Journal of Physics* 35 (1967): 89–98.
4. Leibniz's third paper, February 25, 1716, 9, in Gottfried Leibniz and Samuel Clarke, "Exchange of Papers between Lebiniz and Clarke," http://www.earlymoderntexts.com/assets/pdfs/leibniz1715_1.pdf.
5. Bernhard Riemann, "On the Hypotheses Which Lie at the Foundations of Geometry," trans. William Kingdon Clifford, *Nature* 8 (1873): 14–17, https://www.maths.tcd.ie/pub/HistMath/People/Riemann/Geom/WKCGeom.html.
6. Ibid.
7. Ibid.

4. Einstein, Gravity, and the Universe

1. Edwin F. Taylor and John Archibald Wheeler, *Spacetime Physics: Introduction to Special Relativity* (New York: W. H. Freeman, 1992).
2. Albert Einstein, "The Speed of Light and the Statics of the Gravitational Field," *Annalen der Physik* 38 (1912): 355.
3. Michel Jannssen and Jürgen Renn, "Einstein Was No Lone Genius," *Nature* 527 (2015): 298–300.

4. Albert Einstein, "The Problem of Space, Ether, and the Field in Physics," 1934, in *Ideas and Opinions* (New York: Dell, 1973), 274.

5. Karl Schwarzschild to Albert Einstein, December 22, 1915, in Albert Einstein, *The Collected Papers of Albert Einstein,* vol. 8, *The Berlin Years: 1914–1918* (Princeton, NJ: Princeton University Press, 1998), document 169, http://einsteinpapers.press.princeton.edu/vol8-trans/191.

6. Einstein's model predicts that the distortion of space includes twisting as well as stretching. In this book, I focus just on the stretching of space. For a massive spinning object, such as a black hole, the twisting of space is also important.

7. To be accurate: A person who measures exactly six feet tall atop Mount Everest will also measure exactly six feet tall at sea level. But the two measurements called "six feet" will not be the same length! The yardstick itself will be shorter at sea level by less than the size of an atom.

8. C. W. Chou, D. B. Hume, T. Rosenband, and D. J. Wineland, "Optical Clocks and Relativity," *Science* 329 (2010): 1630–1632.

9. Schwarzschild to Einstein, December 22, 1915.

10. Adrian Cho, "Gravitation Waves, Einstein's Ripples in Spacetime, Spotted for First Time," *Science,* February 11, 2016, http://www .sciencemag.org/news/2016/02/gravitational-waves-einstein-s -ripples-spacetime-spotted-first-time.

11. The lumpiness of galaxies introduces a complication into Einstein's model of gravity. That's because the equations are extremely difficult to solve; in fact, they cannot be solved exactly for a lumpy ("inhomogeneous") universe. Only recently have scientists used supercomputers to approximate solutions to an expanding lumpy universe.

12. There is a cosmic loophole that allows galaxies to recede faster than light. Matter cannot be accelerated through space to the speed of light, because it would take infinite energy to do so. But galaxies are not being accelerated; they are sitting passively in a sea of space and are separating only because new space is welling up between them.

13. This is because there is progressively less matter to distort the space in a given volume. A common misconception is that the force of

gravity slows the expansion. In Einstein's model, gravity is not a force. See Roy R. Gould, "Why Does a Ball Fall? A New Visualization for Einstein's Model of Gravity," *American Journal of Physics* 84 (2016): 96.

14. The confusion arises because the word *scale* is used in two opposite ways. Historically, scientists used the term *scale factor* to describe how much the universe increases in size (for example, by determining the distance between galaxies). The scale factor might better be called the size factor, because that's what it measures.

    By contrast, the scale of the universe is similar to the scale marker on a map. The smaller the scale marker, the larger the area covered by the map. The scale and size are inversely proportional to each other.

15. Another example of this amazing effect: If you could move the Earth into a cube of space, the volume of that cube would increase—even though the boundary of the cube remains fixed! The effect is even greater with a black hole. See Edwin F. Taylor and John Archibald Wheeler, *Exploring Black Holes* (London: Pearson, 2001).

16. Calling it a fudge factor is not really accurate. A fudge factor is an arbitrary parameter that has no physical significance but that is added to an equation to get the desired result. But the cosmological constant was not arbitrary; it was akin to a constant of integration in Newton's laws of motion, for example. The only question was whether its value was zero or something else. Einstein's mistake was not a flaw in his theory; it was a mistake of observation: he simply believed the universe was not changing in scale.

17. Albert Einstein, "Cosmological Considerations in the General Theory of Relativity," *Königlich Preussische Akademie der Wissenschaften,* document 43, February 15, 1917.

18. Svante Arrhenius, "Nobel Prize in Physics 1921—Presentation Speech," December 10, 1922.

19. Alexander Friedmann, letter to Albert Einstein, December 6, 1922, *The Digital Einstein Papers,* vol. 13, document 390, accessed May 20, 2016, http://einsteinpapers.press.princeton.edu/vol13-trans/363.

20. Georges Lemaître, "A Homogeneous Universe of Constant Mass and Increasing Radius Accounting for the Radial Velocity of Extragalactic Nebulae," translated from the original article dated 1925, *Monthly Notices of the Royal Astronomical Society* 91 (1931): 483.

21. Marcia Bartusiak, "Before the Big Bang," *Technology Review,* September / October 2009, M14–M15.

22. "Redshift of Nebulae a Puzzle, Says Einstein," *New York Times,* February 12, 1931, 2.

23. C. O'Raifeartaigh and B. McCann, "Einstein's Cosmic Model of 1931 Revisited: An Analysis and Translation of a Forgotten Model of the Universe," preprint, submitted December 8, 2013, p. 21, https://arxiv.org/pdf/1312.2192.pdf.

24. Ibid., p. 5.

25. Going back in time would be similar to going north on Earth: eventually, when you reach the pole, there is no more north; the very concept of north ends.

5. The Big Bang and Beyond

1. The term *Big Bang* has morphed over the years to include modern insights about the universe's earliest moments, and now the term is ambiguous. In this book, I use *Big Bang* in its original meaning, as it arose from Albert Einstein's model of gravity and Edwin Hubble's astronomical observations. This definition is the one that separates what is known for certain from the more speculative scenarios that extend the Big Bang further back in time.

2. Quoted in "Dr. Albert Einstein Dies in Sleep at 76; World Mourns Loss of Great Scientist," *New York Times,* April 19, 1955, accessed May 20, 2016, http://www.nytimes.com/learning/general/onthisday/bday/0314.html.

3. What was outside that grapefruit-size infant universe? More space. But the region outside the observable part of our universe is beyond what nature allows us to see, so we have no direct evidence of what conditions are like there.

4. The branch of physics called quantum field theory predicts that even empty space is roiling with "virtual" fundamental particles that leap into and out of existence and that contribute energy to space.

5. The law of conservation of energy says only that the amount of energy inside a volume of space must remain constant, unless energy flows into or out of the boundary. When new space comes into existence, it does not disturb the energy inside any other volume of space. It's a good illustration that what we blithely call the "laws of nature" are *our* laws, not nature's—and the laws must be updated as we learn more about nature.

   Furthermore, the universe has a remarkable property: its total energy is zero! That's because the gravitational field has *negative* energy—a unique property not shared by other forces of nature. As a result, the gravitational field of the universe cancels out all the other forms of energy inside itself. By the way, this effect has nothing to do with Einstein; the gravitational field has negative energy even in Isaac Newton's physics!

6. More accurately, they found that the rate of expansion, instead of diminishing, is constant. This causes the size of space to increase exponentially. It is just like having a constant rate of interest in your bank account: your savings grow exponentially even though the interest rate is constant.

7. An analogy in our everyday world is a glowing ember in the fireplace. The atoms in the ember are in a high-energy, unstable state. This energy is transformed into particles of light as the atoms transition to their most stable configuration.

6. Building Plans

1. Physicists envision higher dimensions of space, but stable orbits would require all but three of these space dimensions to be inaccessible.

2. An electron has a magnetic field, like a tiny compass needle, so it can be turned around using a magnet.

3. This stems from a strange property of space. See Roy R. Gould, "Answer to Question #7," *American Journal of Physics* 63 (1995): 109.

4. In the infant universe, neutrons decayed into protons and electrons (the constituents of hydrogen atoms). If neutrons had been lighter, they would not have had enough energy for this decay process. The universe would be filled with neutrons but not atoms or life.

5. B. Carter and W. H. McCrea, "The Anthropic Principle and Its Implications for Biological Evolution," in "The Constants of Physics," special issue, *Philosophical Transactions of the Royal Society of London, Series A, Mathematical and Physical Sciences* 310, no. 1512 (December 20, 1983): 347–363. Also see John D. Barrow and Frank J. Tipler, *The Anthropic Cosmological Principle* (Oxford. Oxford University Press, 1986).

6. N. David Mermin, "QBism Puts the Scientist Back into Science," *Nature* 507 (March 27, 2014): 421–423.

7. J. Robert Oppenheimer, *Uncommon Sense*, ed. N. Metropolis, Gian-Carlo Rota, and David Sharp (Boston: Birkhauser, 1984), 58.

8. Author's personal recollection.

7. An Apple Pie from Scratch

1. National Research Council, *National Science Education Standards* (Washington, DC: National Academy, 1995), 180.

2. Hermann von Helmholtz, "The Thermodynamics of Chemical Processes," in *Wissenschaftliche Abhandlungen* (Leipzig: Barth, 1883), as quoted in Ralph Baierlein and Clayton A. Gearhart, "The Disorder Metaphor," *American Journal of Physics* 71, no. 2 (February 2003): 103. For more on this topic, see Daniel F. Styer, "Entropy and Evolution," *American Journal of Physics* 76, no. 11 (November 2008): 1031–1033; Sean M. Carroll, "The Cosmic Origins of Time's Arrow," *Scientific American* 298, no. 6 (June 2008): 48–57; and Daniel F. Styer, "Insight into Entropy," *American Journal of Physics* 68, no. 12 (December 2000): 1090–1096.

3. George Gamow, *One, Two, Three . . . Infinity: Facts and Speculations of Science* (New York: New American Library / Mentor Books, 1953), 216.

4. Valerie H. Pitt, ed., *The Penguin Dictionary of Physics* (Middlesex, UK: Penguin Books, 1977), 136.

5. National Research Council, *National Science Education Standards,* 180.

6. Elizabeth Garber, Stephen G. Brush, and C. W. F. Everitt, eds., *Maxwell on Heat and Statistical Mechanics: On "Avoiding All Personal Enquiries" of Molecules* (Bethlehem, PA: Lehigh University Press, 1995), 55.

7. Ibid.

8. Ibid., 56.

9. Into the Cauldron

1. John C. Brandt and Ray A. Williamson, "The 1054 Supernova and Native American Rock Art," *Archaeoastronomy* 1 (1979): S1–S38.

2. Cato the Elder, "In Support of the Oppian Law," in *The World's Great Speeches,* ed. Lewis Copeland (Garden City, NY: Garden City, 1942), 32.

3. Ibid., 33.

4. Ibid., 32.

5. Quoted in Helge Kragh, "An Anthropic Myth: Fred Hoyle's Carbon-12 Resonance Level," *Archive for History of the Exact Sciences* 64 (2010): 741.

6. Evgeny Epelbaum et al., "Viability of Carbon-Based Life as a Function of the Light Quark Mass," *Physical Review Letters* 110 (March 15, 2013): 112502.

10. Into the Light

1. Robert E. Blankenship, "Early Evolution of Photosynthesis," *Plant Physiology* 154 (2010): 434–438.

2. Albert Szent-Györgyi, *Introduction to a Submolecular Biology* (New York: Academic Press, 1960), 22.

3. Robert M. Jacobson and Alvan R. Feinstein, "Oxygen as a Cause of Blindness in Premature Infants," *Journal of Clinical Epidemiology* 45, no. 11 (1992): 1265–1287.

4. Gregory J. Retallack, "A 300-Million-Year Record of Atmospheric Carbon Dioxide from Fossil Plant Cuticles," *Nature* 411 (2001): 287–289.

5. Joseph Wood Krutch, *The Great Chain of Life* (Boston: Houghton Mifflin, 1957), 190–191.

6. Andrea Wulf, *The Brother Gardeners* (London: Heinemann, 2008).

7. Galileo, *Selected Writings,* trans. William R. Shea and Mark Davie (Oxford: Oxford University Press, 2012), 181.

8. Ibid.

9. Peter Ward, *Out of Thin Air: Dinosaurs, Birds, and Earth's Ancient Atmosphere* (Washington, DC: Joseph Henry Press, 2006).

10. "The 100 Largest Companies in the World Ranked by Revenue in 2016 (in Billion U.S. Dollars)," Statista, accessed May 25, 2017, https://www.statista.com/statistics/263265/top-companies-in-the-world-by-revenue/.

11. Svante Arrhenius, *Worlds in the Making: The Evolution of the Universe* (New York: Harper and Brothers, 1908), 63.

12. Roger Revelle et al., "Atmospheric Carbon Dioxide," appendix Y4 in *Restoring the Quality of Our Environment: Report of the Environmental Pollution Panel, President's Science Advisory Committee* (Washington, DC: White House, November 1965), 126–127.

13. Department of Defense, *National Security Implications of Climate-Related Risks and a Changing Climate* (Washington, DC: Department of Defense, July 23, 2015), 3. The Pentagon's report goes on to say that "global climate change will have wide-ranging implications for U.S. national security interests over the foreseeable future because it will aggravate existing problems—such as poverty, social tensions, environmental degradation, ineffectual leadership, and weak political institutions." Ibid., 3.

14. William Shakespeare, *A Winter's Tale,* 4.4.

15. Mitch Leslie, "On the Origin of Photosynthesis," *Science* 323 (2009): 1286–1287.

11. The Great Inventor

1. Yung Chih Lai and Cheng-Ming Chuong, "The 'Tao' of Integuments: Hair Follicle and Sweat Gland Fates Can Be Switched by Morphogens at Specific Skin Regions or Developmental Stages," *Science* 354 (2016): 1533.

2. You are part of this process. In fact, every living thing on Earth plays a role in this innovation, in the unfolding of life, and helps to determine the future of life on the planet. Even if you personally do not reproduce, the viruses and microorganisms that live inside you *do* reproduce, mutate, and spread their genes. Your very existence alters the course of evolution.

3. Thomas A. Isenbarger et al., "The Most Conserved Genome Segments for Life Detection on Earth and Other Planets," *Origins of Life and Evolution of Biospheres* 38 (2008): 517–533.

4. The quote was from de Vries's friend Arthur Harris. See Hugo de Vries, *Species and Varieties: Their Origin by Mutation* (Chicago: Open Court, 1904), 826, https://ia902702.us.archive.org/35/items/speciesvarieties00vrieuoft/speciesvarieties00vrieuoft.pdf.

5. Jacques Monod, *Chance and Necessity: An Essay on the Natural Philosophy of Modern Biology* (New York: Vintage Books, 1971), 171–172.

6. Andreas Wagner, *Arrival of the Fittest* (New York: Penguin Random House, 2014), 14.

7. Charles Darwin, *On the Origin of Species* (London: John Murray, 1859), 131.

8. Nicholas Wade, "You Look Familiar. Now Scientists Know Why," *New York Times,* June 1, 2017.

9. Jonathan B. Losos, *Improbable Destinies: Fate, Chance, and the Future of Evolution* (New York: Riverhead Books, 2017), D3.

10. Richard Dawkins, *The Blind Watchmaker* (New York: W. W. Norton, 1986), 139.

12. Information, Please!

1. The sphere was clad in nickel, since plutonium is toxic as well as radioactive.

2. It is the nucleus of the atom that splits.

3. In principle. In practice, it is easier to measure these properties than to calculate them.

4. Michael Baym et al., "Spatiotemporal Microbial Evolution on Antibiotic Landscapes," *Science* 353 (2016): 1147–1151.

5. For the longest-running experiment on bacterial evolution, see B. H. Good, M. J. McDonald, J. E. Barrick, R. E. Lenski, and M. M. Desai, "The Dynamics of Molecular Evolution over 60,000 Generations," *Nature* 551 (2017): 45–50.

6. Gregory J. Chaitin, "Randomness and Mathematical Proof," *Scientific American* 232, no. 5 (May 1975): 47–52.

7. Bruce Alberts, Alexander Johnson, Julian Lewis, Martin Raff, Keith Roberts, and Peter Walter, *Molecular Biology of the Cell*, 4th ed. (New York: Garland Science, 2002), chap. 24.

8. Jacques Monod, *Chance and Necessity: An Essay on the Natural Philosophy of Modern Biology* (New York: Vintage Books, 1971), 125.

9. Ibid., 138.

10. Francis Crick, *Life Itself: Its Origin and Nature* (New York: Simon and Schuster, 1981).

11. Tom Head, ed., *Conversations with Carl Sagan* (Jackson: University Press of Mississippi, 2006), 23.

12. Robert Angus Smith, *Air and Rain* (London: Longmans, Green, 1872).

13. There are nearly an infinite number of protein sequences to explore. That's because the number of ways you can string together letters—whether in a word or in a protein—becomes huge even with a small number of letters. For example, the word *infrastructure* has a mere fourteen letters, yet there are more than 60,000,000,000,000,000,000 different fourteen-letter words that are possible using the English alphabet. (Two random examples are

lmqqdoreipscas and *iuwiojclawehsc*—hardly words to find inspiring.) For proteins, which have more "letters," the number of possible combinations is vastly greater.

14. Gerald F. Joyce, "The Counterforce," *Current Biology* 9 (1999): R500.

15. Gerald F. Joyce, "Reflections of a Darwinian Engineer," *Journal of Molecular Evolution* 81 (2015): 146.

16. That's because RNA has but a single strand, which can fold into a complex, three-dimensional shape. DNA is double-stranded, so it doesn't fold as readily.

17. Gerald F. Joyce, "The Antiquity of RNA-Based Evolution," *Nature* 418 (2002): 214–220.

18. Tracey A. Lincoln and Gerald F. Joyce, "Self-Sustained Replication of an RNA Enzyme," *Science* 323 (2009): 1229–1232, http://doi.org /10.1126/science.1167856.

19. Charles Olea Jr., David P. Horning, and Gerald F. Joyce, "Ligand-Dependent Exponential Amplification of a Self-Replicating L-RNA Enzyme," *Journal of the American Chemical Society* 134, no. 19 (2012): 8050–8053, doi:10.1021/ja302197x.

20. Joyce, "The Counterforce," R501.

21. Peter D. Ward and Donald Brownlee, *Rare Earth: Why Complex Life Is Uncommon in the Universe* (New York: Copernicus, 2000).

13. Is Evolution Predictable?

1. Hermann Klencke and Gustav Schlesier, *Lives of the Brothers Humboldt: Alexander and William,* trans. Juliette Bauer (London: Ingram, Cooke, 1852), 226–230. All quotes in the account given here are from this source.

2. Charles Darwin, *On the Origin of Species* (London: John Murray, 1859), 193.

3. Ibid., 193–194.

4. Jason R. Gallant et al., "Genomic Basis for the Convergent Evolution of Electric Organs," *Science* 344 (2014): 1522–1525.

5. Shaohua Fan et al., "Going Global by Adapting Local: A Review of Recent Human Adaptation," *Science* 354 (2016): 54–58.

6. M. Sabrina Pankey et al., "Predictable Transcriptome Evolution in the Convergent and Complex Bioluminescent Organs of Squid," *Proceedings of the National Academy of Sciences* 111, no. 44 (2014): E4736–E4742, http://www.pnas.org/cgi/doi/10.1073/pnas.1416574111.

7. Caroline Ash, "How Fish Evolved an Abyssal Glow," *Science* 353 (2016): 360. See also Matthew P. Davis, John S. Sparks, and W. Leo Smith, "Repeated and Widespread Evolution of Bioluminescence in Marine Fishes," *PLOS ONE* 11, no. 6 (2016): e0155154, doi:10.1371/journal.pone.0155154.

8. Neal Anthwal, Leena Joshi, and Abigail S. Tucker, "Evolution of the Mammalian Middle Ear and Jaw: Adaptations and Novel Structures," *Journal of Anatomy* 222, no. 1 (2013): 147–160, http://doi.org/10.1111/j.1469-7580.2012.01526.x. See also University of Chicago Medical Center, "Prehistoric Jawbone Reveals Evolution Repeating Itself," *Science News,* February 16, 2005, https://www.sciencedaily.com/releases/2005/02/050212203647.htm.

9. Ronald R. Hoy, "Convergent Evolution of Hearing," *Science* 338 (2012): 894–895. See also Fernando Montealegre-Z. et al., "Convergent Evolution between Insect and Mammalian Audition," *Science* 338 (2012): 968.

10. Ping Wu et al., "Molecular Shaping of the Beak," *Science* 305 (2004): 1465–1466, http://doi.org/10.1126/science.1098109. Also see Arhat Abzhanov et al., "*Bmp4* and Morphological Variation of Beaks in Darwin's Finches," *Science* 305 (2004): 1462–1464.

11. Robert E. Shadwick, "How Tunas and Lamnid Sharks Swim: An Evolutionary Convergence," *American Scientist* 93 (2005): 524.

12. David W. Murphy et al., "Underwater Flight by the Planktonic Sea Butterfly," *Journal of Experimental Biology* 219 (2016): 535–543.

13. Matthew C. Brandley, John P. Huelsenbeck, and John J. Wiens, "Rates and Patterns in the Evolution of Snake-Like Body Form in Squamate Reptiles: Evidence for Repeated Re-evolution of Lost Digits and Long-Term Persistence of Intermediate Body Forms,"

Evolution 62, no. 8 (August 2008): 2042–2064, doi:10.1111/j.1558 -5646.2008.00430.x.

14. Susan Evans, "Four Legs Too Many?," *Science* 349 (2015): 374–375. See also David M. Martill, Helmut Tischlinger, and Nicholas R. Longrich, "A Four-Legged Snake from the Early Cretaceous of Gondwana," *Science* 349 (2015): 416–419.

15. C. Tristan Stayton, "What Does Convergent Evolution Mean? The Interpretation of Convergence and Its Implications in the Search for Limits to Evolution," *Interface Focus* 5, no. 6 (December 6, 2015): 20150039.

16. Jamie T. Bridgham, "Predicting the Basis of Convergent Evolution," *Science* 354 (2016): 289; David L. Stern and Virginie Orgogozo, "The Loci of Evolution: How Predictable Is Genetic Evolution?," *Evolution* 62, no. 9 (2008): 2155–2177; David L. Stern, "The Genetic Causes of Convergent Evolution," *Nature Reviews Genetics* 14 (2013): 751–764.

17. Angela M. Hancock, "How Conifers Adapt to the Cold," *Science* 353 (2016): 1362–1363. Also see Sam Yeaman et al., "Convergent Local Adaptation to Climate in Distantly Related Conifers," *Science* 353 (2016): 1431–1433.

18. Andreas Wagner, *Arrival of the Fittest: How Nature Innovates* (New York: Penguin Random House, 2014).

19. George McGhee, *Convergent Evolution: Limited Forms Most Beautiful* (Cambridge, MA: MIT Press, 2011). See also Jonathan B. Losos, *Improbable Destinies: Fate, Chance, and the Future of Evolution* (New York: Riverhead Books, 2017).

14. The Sensational Sensations

1. James Agee, "Knoxville: Summer of 1915," *Partisan Review,* August– September 1938, 22–25.

2. For example, see Frank Jackson, "Epiphenomenal Qualia," *Philosophical Quarterly* 32 (1982): 127–136.

3. And this is true of the other senses, not just vision. For example, there is a component of male perspiration that males supposedly do

not sense but women do. Half of women find the aroma pleasant, while the rest find it repulsive. I confess that I did not believe this until a colleague bought a bottle of the compound and opened it in the office. I heard women crying up and down the corridor, "What's that?" and "Ughh. What's *that?*" The men had no idea what the women were talking about.

4. Quoted in Joseph Wood Krutch, *The Great Chain of Life* (Boston: Houghton Mifflin, 1957), 202–203.

5. Ibid., 203. See also J. Robert Oppenheimer, "Uncommon Sense," lecture 5, *The Reith Lectures,* BBC Radio, December 13, 1953.

6. Andrei Linde, "Universe, Life, Consciousness," *Science and Nonduality,* 2015, https://www.scienceandnonduality.com/wp-content/uploads/2015/11/UNIVERSE-LIFE-CONSCIOUSNESS-Andrei-Linde.pdf.

15. Design without a Designer?

1. Here's how to determine the color (or gray shade) of each point on your canvas.

Start with the coordinates of the point. Let's say the point is at (2, 1). This is your **original point**, the one you are working on.

Now make two numbers, as follows:

For the first number:
Square the x coordinate.
Square the y coordinate and subtract it from the first result.
Then add one.
The result is your first number.

For the second number:
Multiply the x coordinate by the y coordinate.
Multiply the result by two.
The result is your second number.

These two numbers form the coordinates of a *new* point.

If the distance from the new point to the origin (0,0) is *greater* than two units, color the **original point** any color (or gray) of your choice.

If the distance from the new point to the origin (0,0) is *less* than two units, repeat the process, using the new set of numbers rather than the original coordinates. That is, square the first number, square the second number and subtract, and so on.

If the distance from the *new* new point to the origin (0,0) is greater than two units, color the **original point** that you started with a slightly different color or gray shade.

Otherwise, repeat the process.

Keep repeating the process until you get a new point that is farther than two units from the origin. Then color your original point.

The color of your original point depends on how many times you have to repeat the process in order to first get a point that is farther than two units from the origin.

Repeat the preceding steps for every point on your canvas.

For each point on your canvas, the number of iterations (repeats) you have to make to get beyond two determines the color (or gray shade) of the point. For example, two iterations might be blue, and three iterations might be blue-green, and four iterations might be green. Always use the same color for the same number of iterations.

2. This is a hallmark of chaotic behavior. It limits our ability to predict how something will turn out based on what we already know about it.

16. "Who's There?"

1. Courtney D. Dressing and David Charbonneau, "The Occurrence Rate of Small Planets around Small Stars," *Astrophysical Journal* 767 (2013): 95.

2. Pluto has since been demoted to a "dwarf planet," though not all astronomers agree with the reclassification.

3. The stars are named after the projects that search for them, such as the Transatlantic Exoplanet Survey (TRES).

4. Daniel Clery, "Forbidden Planets," *Science* 353 (July 29, 2016): 438–441.

5. Dressing and Charbonneau, "Occurrence Rate of Small Planets," 95.

6. A trillion times fainter because the illumination from a source of light decreases as the square of its distance from you.

7. If the surface were liquid, there would be tides a hundred feet high.

8. J. Hunter Waite et al., "Cassini Finds Molecular Hydrogen in the Enceladus Plume: Evidence for Hydrothermal Processes," *Science* 356 (2017): 155–159.

9. Daniel Clery, "Signs of Life," *Science* 358 (November 3, 2017): 578–581.

10. Sara Seager, "The Future of Spectroscopic Life Detection on Exoplanets," *Proceedings of the National Academy of Sciences* 111, no. 35 (2014): 12634–12640.

11. Abraham Loeb, Rafael Batista, and David Sloan, "Relative Likelihood for Life as a Function of Cosmic Time," preprint, submitted June 27, 2016, https://arxiv.org/abs/1606.08448.

Acknowledgments

I am grateful to Kenneth Bergman, Avi Loeb, Jean-David Rochaix, Irwin Shapiro, and anonymous reviewers for their comments and suggestions about the material in this book. Special thanks to Kim Kronenberg and Allen Taylor for their advice and encouragement. Any errors of fact are solely mine.

At Harvard University Press, I thank Emeralde Jensen-Roberts for her keen eye and judgment, and especially my editor, Jeff Dean, for his gentle guidance and critical insights.

Portions of the material in this book were inspired by a workshop that I led for the Teachers as Scholars program, developed by Henry Bolter and hosted at Harvard University and Tufts University.

Index

Alpher, Ralph, 71–72
Anasazi people, 111
Andromeda galaxy, 33–34, 39
Antibiotic resistance, 166–168
Arrhenius, Svante, 65–66, 135–136,
    178; on Einstein, 65–66; on global
    warming, 135–136; on panspermia
    theory, 178

Banks, Joseph, 25–26, 133
Big Bang, 2–3, 6, 12, 36, 70–74,
    78–82, 84, 105, 240–241; as
    boundary of ignorance, 79;
    evidence for, 71–73; as
    transition, 79
Boltzmann, Ludwig, 99
Bonpland, Aimé, 185
Building plan, 3–13, 83–92, 144;
    fine-tuning, 9–10, 87–90; forces of
    nature, 83; number of dimensions,
    85; properties of atoms, 84–87;
    quantum rules, 86

Cairns, John, 138
Caroline of Ansbach, 44–45, 47
Cato the Elder, 115
Chaitin, Gregory, 168–170, 172
Chance. *See* Randomness
Chemical elements, 3, 12, 71, 74, 84,
    87, 105, 120, 122; carbon,
    116–118; creation in stars, 112–118
Chemical equilibrium, 131–134
Chinese astronomers, 111
Clarke, Arthur C., 229–230
Clarke, Samuel, 45
Clausius, Rudolph, 102
Climate change, 134–137
Consciousness, 9, 125, 209–212
Convergent evolution, 185–203;
    hearing, 190–192; *Leuchtenbergia
    principis,* 193–195; locomotion,
    195–197; mating behavior,
    197–199; sea snails, 196–197;
    shape-forming proteins, 193;
    sharks and tunas, 195–196;

Convergent evolution (*continued*)
significance of, 199–202; squid
luminescence, 189–190. *See also*
Evolution of life
Cook, James, 25–26
Corals, 130–131
Cosmological constant, 65–68
Crick, Francis, 177
Curtis, Heber, 28, 31–33

Darwin, Charles, 2, 54, 124,
157–159, 181–182, 187, 193
D'Auteroche, Jean Chappe, 23
Dawkins, Richard, 161
De Duve, Christian, 10
De Vries, Hugo, 157
DNA, 96, 152, 154–155, 187, 242;
and genes, 147–150; mutation in,
158, 200–202; and origin of life,
164, 177, 181–182; and ultraviolet
light, 129, 235
Drake, Frank, 223

Einstein, Albert, 5, 29, 32, 39, 43,
91, 211, 243; model of gravity,
53–59, 74–76, 78–80; model of
universe, 59–63; and skepticism
about expanding universe, 64–69;
solitude, 73; and special relativity,
52–53
Electric eel, 185–188
Enceladus, 231
Entropy, 96–103, 183; apparent
paradox, 96–99; and forces of
nature, 100–101; and heat death,
101–102
Europa, 228–232

European Space Agency (ESA), 73,
107
Evolution of life, 8, 54, 121, 144,
146–147, 157, 222, 237; and
antibiotic resistance, 166–168; and
arrival of the fittest, 157, 159; and
chance, 8–9, 158–159, 167, 176,
183–184; and Darwin, 54, 157,
181–182, 187; and innovation,
157–159; laboratory-based, 162,
178–183; language of, 160; and
lifetime of the Sun, 121; and
machinery of life, 146–147, 159;
pathways of, 144, 199–202; and
predictability, 166–168, 185–212,
222; and role of time, 161–162.
See also Convergent evolution
Exoplanets, 12–13; around dwarf
stars, 234–235; detection of,
225–228; diversity of, 223–225;
and Earth, 235–236; evidence for
life on, 232–235; and intelligent
life, 236–238
Expanding universe: accelerating,
74–77; counterintuitive, 64–65;
and creation of space, 62; and dark
energy, 75–76; Einstein's skepticism
about, 65–68; expands inward,
63–64; fine-tuning of, 77; and
formation of structure, 106; and
Friedmann, 66–68; and Lemaître,
67; observations of, 39–42, 76–77;
problems with, 74; size of effect,
62–63. *See also* Universe

Fermi, Enrico, 236
Fermi paradox, 236–238

Fine-tuning, 9–10, 87–90, 117–119; and carbon, 117–119; defined, 9–10; perspectives on, 88–92

Friedmann, Alexander, 66–68, 71

Galaxies, 5, 13, 27–34, 37; Andromeda, 33–34, 39; distance to, 33, 62, 67; and exoplanets, 222, 225, 236; formation of, 97, 110; Milky Way, 27, 28, 30–31; motions of, 32, 33, 37–41, 61–62, 67; number of, 32, 241; size of, 30–31

Galileo, 27, 133–134

Gamow, George, 70–71, 99, 210

Gould, Stephen J., 8

Gravitational waves, 58

Great Debate, 28–35

Green, Charles, 25–26

Halley, Edmond, 20–22

Hawking, Stephen, 14

Helios, 19, 239

Helmholtz, Hermann von, 99, 102

Herman, Robert, 71–72

Hoyle, Fred, 72, 117–118, 178; and carbon formation, 117–118; derides "Big Bang," 72; and panspermia theory, 178

Hubble, Edwin, 39–42, 61–62, 68–69, 72

Hubble Space Telescope, 13, 228, 234

Humason, Milton, 39, 68

Humboldt, Alexander von, 185–187

Immune system, 174–177

Indian pipe, 131

Inflationary universe, 79–81

James Webb Space Telescope, 234

Joyce, Gerald F., 181–183

Julesz, Béla, 207

Jupiter, 223, 224, 228, 230, 232

Kelvin, Lord (William Thomson), 102

Le Gentil, Jean Baptiste, 24–25

Leibniz, Gottfried, 44–47, 52, 57, 88

Lemaître, Georges, 67

Leuchtenbergia principis, 193–195

Life, 3, 7–13, 77, 84–92, 103, 108, 116–121, 123; carbon-based, 116–119; and chemical equilibrium, 103, 131–134, 139–140; and climate change, 134–137; and convergent evolution, 188–203; and creativity, 137–139; diversity of, 153–155; as innovator, 126, 143–144; intelligent beings, 236–238; machinery of, 146–150; and natural selection, 157–161; origins of, 177–183; on other planets, 222–238; and oxygen, 128–129; and physical sensations, 204–212; potential evidence for, 232–235; powered by sunlight, 124–140; and properties of stars, 119–121, 123; and randomness, 164, 166–168, 174–177; re-use of genes, 192–193; robustness of, 150–153; ubiquity of, 155–156; and universe's building plan, 7–13, 77, 84–92, 103, 108, 144. *See also* Convergent evolution; DNA; Evolution of life

Light: as carrier of information, 13, 36, 232–234, 241; constant speed of, 51–52; and relativity, 52
LIGO, 58
Limacina helicina, 196–197
Linde, Andrei, 9, 211–212, 244
Lord Howe Island, 197
Lorentz, Hendrik, 52

Mandelbrot, Benoit, 213
Mandelbrot set, 213–221
Maudlin, Tim, 10
Maxwell, James Clerk, 54
Monod, Jacques, 4, 8, 158, 175–176
Morrison, Philip, 177
Multiverse, 79–81, 89–90

NASA, 73, 107, 224, 226, 230–232, 234
Newton, Isaac, 44–49, 52–56, 88, 211
Newton-Leibniz debate, 44–47
Noble, Tim, 244
Nuclear fusion, 112–114, 117–120, 122

Oppenheimer, J. Robert, 210

Penzias, Arno, 72
Photosynthesis, 125–131; and climate change, 134–137; problem of oxygen, 128–129; variations on, 129–131
Proteins, 154–155, 178–180, 192–193

Randomness, 80, 145, 159, 164, 165–166, 167–173, 176; and

evolution of proteins, 178–180; and evolution of RNA, 181–183; and formation of structure, 107; and information, 168–174; and predictability, 165–168
Riemann, Bernhard, 47–49, 53, 55, 57, 71, 219
RNA: as carrier of information, 148–149; and origin of life, 164, 180–183

Sagan, Carl, 102, 140
Scale of length, 45–50, 52, 57–65, 67, 81, 216–219
Schwarzschild, Karl, 56–57
Sensations, 204–212; color, 206–208, 209; and consciousness, 209–212; as emergent phenomenon, 208–209
Shakespeare, William, 3, 6, 43–44, 138–139
Shapley, Harlow, 28–33, 35
Significance of human life, 35–36, 137–139, 243–245
Slipher, Vesto, 39
Smith, Robert A., 178
Solar system, 12, 20, 26, 39, 223–224, 228, 237; compared to atoms, 85; and distance ladder, 30, 226; location of, 31, 35; scale of, 20–21, 27, 226
Space: expands inward, 63–64; mass of, 75–77; Newton's and Leibniz's views, 44–47; Riemann's view, 47–50; scale of distance, 45–50, 60–62; singularity in, 69
Squid, 189–190

Stars: and building plan, 117; carbon formation, 116–119; and chemistry, 123; creation of chemical elements, 112–118; dwarf stars, 234–235; explosion of, 111, 114–116, 119–120; as incubators, 121–122; and photosynthesis, 125–126

Sun, distance and size of, 19–26

Sunflowers, 130

Szent-Györgyi, Albert, 126

Snootak, Jack, 179 180

Thomson, William (Lord Kelvin), 102

Transit of Venus, 20–26; seen from India, 24–25; seen from Mexico, 23–24; seen from Tahiti, 25–26; significance, 26–27

Twain, Mark, 4–5

Tyrian purple, 115

Universe: and Big Bang, 70–79; building plan, 83–92; common conceptions, 2–7; and creativity, 137–138; as entity in itself, 5–7; and entropy, 96–104; expansion

of, 39–42; and forces of nature, 103; formation of structure, 97–108; as infant, 96–98, 105–110; as innovator, 143–147, 153–155; multiple universes, 79–81; mystery, 15; our place in, 35–36; still in creation, 2–3, 42; uses waste heat, 103–104, 112. See also Building plan; Expanding universe; Significance of human life

Wagner, Andreas, 158 159

Water: and climate change, 137; and expanding universe, 105, 108; and formation of order, 101; importance for life, 13, 77, 87, 96, 144; and photosynthesis, 125–129, 139

Webster, Sue, 244

Weinberg, Steven, 8, 40

Wheeler, John A., 7–13, 164, 222, 238; cosmos and consciousness, 90–92, 212, 220, 244; guarantee of life?, 7–12; his question rephrased, 173; and search for life, 243

Whitman, Walt, 13, 126

Wilson, Robert, 72

Woodward, Robert, 126